The Circulatory & Respiratory Systems

Titles in the series

The Muscular & Skeletal Systems

The Circulatory & Respiratory Systems

The Digestive & Urinary Systems

The Reproductive & Nervous Systems

THE HUMAN BODY
& THE ENVIRONMENT
How our surroundings affect our health

The Circulatory & Respiratory Systems

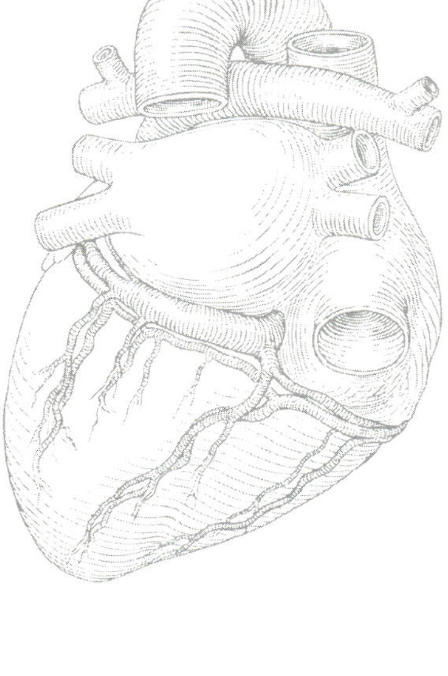

GREENWOOD PRESS
Westport, Connecticut · London

Library of Congress Cataloging-in-Publication Data

Creative Media Applications
The human body & the environment/how our surroundings affect our health
p. cm.—(Middle school reference)
Includes bibliographical references and index.
Contents: v. 1. The muscular & skeletal systems—v. 2. The circulatory & respiratory systems—v. 3. The digestive & urinary systems—v. 4. The reproductive & nervous systems.
ISBN 0-313-32558-8 (set: alk. paper)—0-313-32559-6 (v.1)—0-313-32560-X (v.2)—0-313-32561-8 (v.3)—0-313-32562-6 (v.4)
1. Human physiology—Juvenile literature. [1. Human physiology. 2. Environmental health.] I. Title: Human body and environment. II. Series.
QP37.H8925 2003
612—dc21 2002035217

British Library Cataloguing in Publication Data is available.

Library of Congress Catalog Card Number: 2002035217

ISBN: 0–313–32558–8 (set)
0–313–32559–6 (vol. 1)
0–313–32560–X (vol. 2)
0–313–32561–8 (vol. 3)
0–313–32562–6 (vol. 4)

First published in 2003

Greenwood Press, 88 Post Road West, Westport, CT 06881
An imprint of Greenwood Publishing Group, Inc.
www.greenwood.com

Printed in the United States of America

The paper used in this book complies with the Permanent Paper Standard issued by the National Information Standards Organization (Z39.48–1984).

10 9 8 7 6 5 4 3 2 1

A Creative Media Applications, Inc. Production
WRITER: Robin Doak
DESIGN AND PRODUCTION: Fabia Wargin Design, Inc.
EDITOR: Matt Levine
COPYEDITOR: Laurie Lieb
PROOFREADER: Barbara Francis
ASSOCIATED PRESS PHOTO RESEARCHER: Yvette Reyes
CONSULTANT: Michael Windelspecht

PHOTO CREDITS:
Cover: (*top*) © Craig Aurness/CORBIS and (*bottom*) © S. P. Gillette/CORBIS
AP/Wide World Photographs *pages* vi, ix, 6, 8, 16, 24, 33, 35, 42, 48, 56, 59, 62, 64, 68, 72, 78, 81, 82, 85, 88, 94, 96, 97, 98, 99, 100, 105, 106, 107, 108, 112, 114, 116, 118, 119, 121, 123
© Bettmann/CORBIS *pages* 10, 58, 125
© Photodisk *pages* 14, 28, 66, 110, 117
© Hossler, Ph.D./Custom Medical Stock Photo *page* 19
© Keith/Custom Medical Stock Photo *page* 23
© Howard Sochurek/CORBIS *page* 27
© M. Stocklein/Custom Medical Stock Photo *page* 30
© T.Bannor/Custom Medical Stock Photo *page* 36
© L.Birmingham/Custom Medical Stock Photo *pages* 39, 54
© AFP/CORBIS *page* 52
© Logical Images/Custom Medical Stock Photo *page* 60
© P.Berndt/Custom Medical Stock Photo *page* 70
© M.Bryson/Custom Medical Stock Photo *page* 90
© K.Beebe/Custom Medical Stock Photo *page* 91
© Science Pictures Limited/CORBIS *page* 93

ILLUSTRATION CREDITS:
© Lifeart *pages* iii, 1, 2, 4, 5, 11, 55, 74, 87, 89, 95

CONTENTS

INTRODUCTION

LIVING IN A RISKY ENVIRONMENT

Each and every day, people come in contact with environmental agents that can impact their health. Sunlight, chemicals, air pollution, and bacteria are just a few of the environmental agents that affect the way the human body functions. These agents may be present in the air we breathe or the food and water we ingest.

Environmental health is the human body's reaction to environmental agents, including both natural and human-made substances. In many cases, people have little or no control over their amount of exposure to these environmental agents. In other cases, people choose to do things that can cause or worsen medical problems. Smoking cigarettes and drinking alcohol, for example, can negatively affect a person's health. Lifestyle decisions made by one person can also affect other people. For example, secondhand cigarette smoke has been proven to have harmful effects on the respiratory systems of children and adults who are exposed to it.

Just as some lifestyle decisions contribute to ill health, other decisions help keep the body as healthy as possible. Eating a proper diet and exercising are two important steps that you can take to stay physically fit. Another part of the picture is learning how environmental conditions affect us and how to minimize the damage from environmental agents.

Opposite:
Heart transplant patient Kelly Perkins, age 36, makes her way up Mount Whitney in California. Just two years after her heart surgery, Perkins put her new heart to an amazing test by climbing the highest peak in the United States outside of Alaska.

THE ENVIRONMENT AND THE BODY

Like other body systems, the circulatory system can be seriously affected by the environment. Radiation, chemical exposure, and biological agents can all cause damage to the heart and other parts of the circulatory system. Smoking is especially harmful to this body system. It has been linked to high blood pressure, heart disease, blood clots, and strokes. Diets that are high in cholesterol can also damage the heart.

Everything that we inhale has the potential to affect the body's respiratory system. Air pollution, chemicals, viruses, and bacteria can all harm the lungs and cause serious respiratory conditions. Pollution and chemicals, for example, have been linked to irritated lungs, asthma, sinusitis, and cancer. Smoking is particularly harmful to the respiratory system. It has been shown to worsen existing respiratory conditions and has also been shown to cause cancer, emphysema, and other diseases. Cigarette smoke, which contains numerous *carcinogens* (kar-SIN-uh-jenz), or things that cause cancer, can lead to death. Each year, smoking causes the deaths of more than 400,000 American men and women; secondhand smoke causes about 3,000 deaths.

The immune system is the body system that fights viruses, bacteria, and other organisms that cause illness and disease. In some cases, the immune system does not function properly and attacks normally harmless substances from the environment that get into the body, such as dust, pollen, or pet dander. This is called an *allergy.* One virus that targets, attacks, and damages the immune system is HIV. HIV can lead to AIDS, a serious, life-threatening condition.

HOW TO USE THIS BOOK

Each volume of this series covers a number of body systems and the diseases—environmental and otherwise—that can affect them. Each volume is organized to include the basic anatomy of each system, including its structure and function. This volume includes the circulatory, respiratory, and immune systems.

Opposite:
A high school student takes a smoking break outside Findlay High School in Findlay, Ohio. In many states it is illegal for minors to purchase or possess tobacco products because of their adverse health consequences.

Non-environmental medical conditions and environmental medical conditions for the body systems are covered in separate chapters. Each disease listing includes the condition's causes, symptoms, and treatments, and cures. Additional information may include diagnostic tools, statistics, and historical information.

The final chapter in each volume offers tips on how to keep the body healthy. The glossary gives definitions and pronunciations of some words in the book. For a complete list of pronunciations, however, consult a medical dictionary. The index can be used to find a particular disorder easily.

Finally, the volumes are not meant to be used for self-diagnosis of medical conditions. Those who have health problems should always consult a medical professional.

Note: All metric conversions in this book are approximate.

1

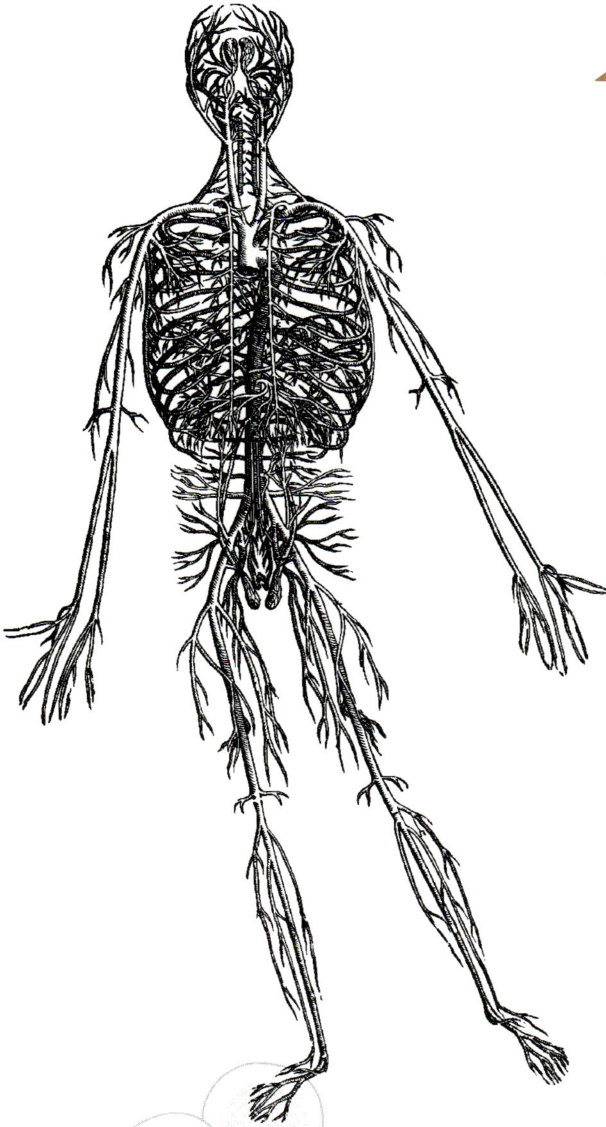

THE CIRCULATORY SYSTEM

To stay healthy, the human body needs a constant supply of oxygen and nutrients. Waste matter, including carbon dioxide, must be removed. The *circulatory* (SURK-yoo-luh-tor-ee) *system* performs these important tasks. Also known as the cardiovascular (kar-dee-oh-VAS-kyoo-lur) system, the circulatory system is made up of the heart, blood vessels, and blood.

THE HEART

The heart is the organ that powers the circulatory system. Every minute of every day, the heart is working automatically. It never stops to rest. This strong, fist-sized muscle consists of two side-by-side pumps. The right side pumps blood into the lungs to be *oxygenated,* or supplied with oxygen. The left side pumps oxygenated blood throughout the body, to be used by organs, cells, and tissues.

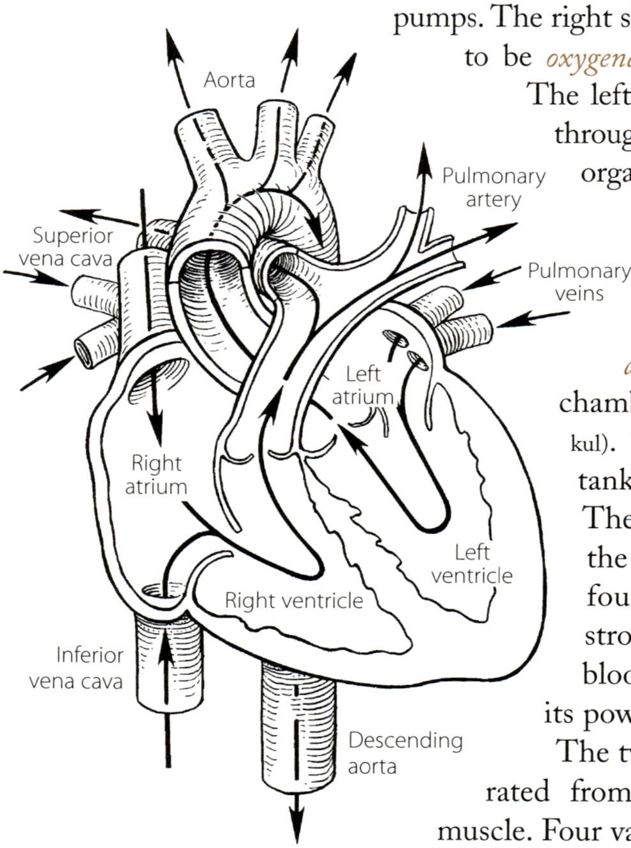

The heart is made up of four different chambers. Each side of the heart has an upper chamber, called an *atrium* (AY-tree-um), and a lower chamber, called a *ventricle* (VEN-trih-kul). The atria serve as holding tanks for blood entering the heart. The ventricles are responsible for the heart's pumping action. All four chambers are lined with strong cardiac muscle that pushes blood throughout the body with its powerful contractions.

The two sides of the heart are separated from one another by a wall of muscle. Four valves control the flow of blood into and out of the heart. The valves open and close, ensuring that blood flows in one direction only.

The heart is well protected within the body. In the chest cavity, this important organ is shielded by the ribs. The heart itself is covered with a protective membrane called the *pericardium* (perr-ih-KAR-dee-um). The space between the pericardium and the heart is filled with fluid.

Like other effective machines, all the parts of the human body must work together. The circulatory system works together with the respiratory system to get oxygen from the lungs to the rest of the body. It delivers carbon dioxide and other waste matter to the lungs, which is removed when a person exhales. The circulatory system works with the digestive system to absorb nutrients from foods that have been ingested. It also works with the excretory system, filtering blood through the kidneys to remove waste products.

Pulmonary and Systemic Circulation

There are two types of circulation in the body: pulmonary and systemic. As these two types of circulation work together, they produce a distinctive sound. This "lub-dub" sound is more commonly known as a heartbeat. The sound is made as the valves of the heart open and close, allowing blood to flow into and out of the heart.

- *Pulmonary circulation* takes place on the right side of the heart. It occurs when the right ventricle pumps oxygen-poor blood into the lungs. In the lungs, the blood takes in oxygen and is returned to the left side of the heart.
- *Systemic circulation* takes place on the left side of the heart. It occurs when the left ventricle pumps oxygen-rich blood to the rest of the body and returns deoxygenated blood to the right side of the heart.

BLOOD VESSELS

The blood vessels are an intricate network of tubes that carry blood to every organ and area of the body. There are three main types of blood vessels in the body: arteries, veins, and capillaries. Blood circulates through these vessels in a continuous cycle throughout the body.

Arteries are the strong, elastic tubes that carry blood away from the heart. Smaller blood vessels, called *arterioles,* branch off from the arteries. Because the blood that pulses through the arteries is under great pressure, these blood vessels have to be thick and strong.

Veins carry blood from the body to the heart. Smaller blood vessels, called *venules,* branch out from the veins. Because veins are less active and carry blood that is under less pressure, they don't have to be as strong as arteries. There are two types of veins: *Superficial veins* lie near the surface of the skin. *Deep veins* are the big veins in the muscles and other parts of the inner body.

Capillaries (KAP-ih-leh-reez) are tiny blood vessels that link arteries and veins. They are found in the tissues throughout the body. Capillary walls are very thin and allow oxygen, carbon dioxide, and other matter to enter and exit easily. Nutrients and waste are constantly being exchanged in the capillaries.

Big Blood Vessels

The following are important blood vessels in the body.

- The *aorta* (ay-OR-tuh) is the largest blood vessel in the entire body. This large artery measures about 1 inch (2.5 centimeters) in diameter. The aorta's job is to carry oxygen-rich blood from the left ventricle to the other arteries in the body.
- The *pulmonary artery* carries oxygen-poor blood from the right ventricle to the lungs.
- The *venae cavae* (VEE-nee CAY-vee) are the two largest veins in the body. The superior vena cava, located above the heart, carries blood from the head, neck, and arms to the right atrium. The inferior vena cava, located below the heart, carries blood from other parts of the body to the right atrium.
- The *carotid* (kuh-RAH-tid) *artery* carries oxygen-rich blood from the left ventricle to the head.
- The *jugular* (JUG-yoo-lur) *vein* carries oxygen-poor blood from the head to the right atrium.

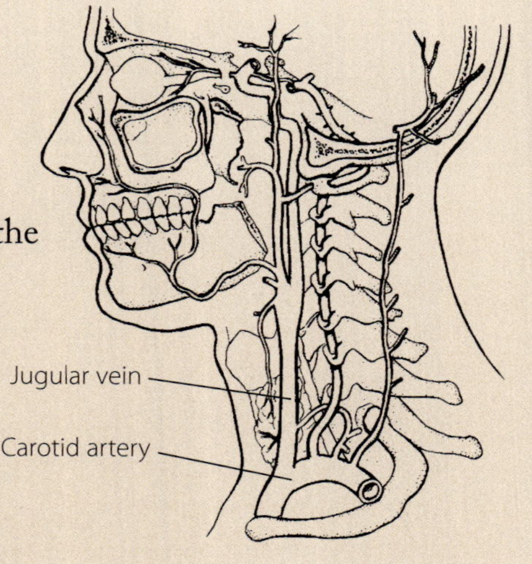

BLOOD

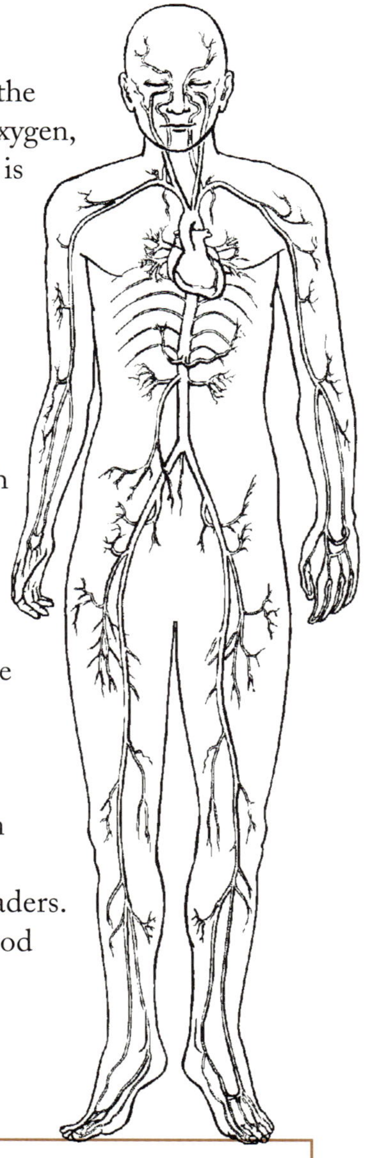

The heart powers circulation. Blood vessels are the pathways. However, a fluid is needed to transport oxygen, nutrients, and waste to where they need to go. This is the function of blood.

As blood circulates throughout the body, it provides cells and tissues with oxygen, protein, carbohydrates, and other nutrients. At times, it may also deliver substances around the body that are less healthy, including bacteria, viruses, and toxins. Blood also removes waste products from cells and tissues and delivers them to organs, such as the lungs and kidneys, that are responsible for waste removal.

Blood also helps regulate body temperature. When the body heats up, blood vessels in the skin *dilate,* or widen, releasing heat and cooling the body. The opposite occurs when the body is cold, conserving heat.

Blood plays an important role in the immune system, too. The *immune system* fights infection in the body. Certain types of blood cells defend the body from bacteria, viruses, and other foreign invaders. Blood also keeps the body healthy by *clotting.* Blood cells clot, or clump together, at a wound site and form a scab, which controls bleeding. Clotting ensures that wounds heal quickly and properly.

Blood is a watery fluid that contains many different types of cells. When blood contains oxygen, it is bright red in color. When it lacks oxygen, it is a darker red. Elements that make up blood include plasma, red blood cells, white blood cells, and platelets.

Fast Fact

Medical professionals take blood from veins instead of arteries, because it is safer and easier. The blood in veins is under less pressure and flows more slowly. As a result, it doesn't gush out and cause excessive blood loss.

Plasma is a clear fluid that is about 90 percent water. More than half of blood is made up of plasma. In addition to carrying blood cells, plasma also contains dissolved nutrients, minerals, and proteins—substances needed by the body's tissues.

Also called erythrocytes (uh-RITH-roh-sites), *red blood cells* are the most numerous of all blood cells. There are about 1 billion red blood cells in every two to three drops of blood. Red blood cells look like round disks that have been flattened in the center. A red pigment in these cells, called *hemoglobin* (HEE-moh-gloh-bin), gives blood its color. Red blood cells carry oxygen from the lungs to the body's tissues. They also collect carbon dioxide and other gases to carry back to the lungs. This type of blood cell is produced in the red bone marrow of some bones.

Also called leukocytes, *white blood cells* play an important role in the immune system. They fight infections and germs. White blood cells are much less plentiful than red blood cells. For every white blood cell in the body, there are 600 red blood cells.

Platelets are the smallest type of blood cell. They contribute to blood clotting by adhering to the lining of blood vessels. Blood clotting is the first step in the healing of any wound. Without clotting at a wound site, blood would continue to flow freely, resulting in excessive blood loss.

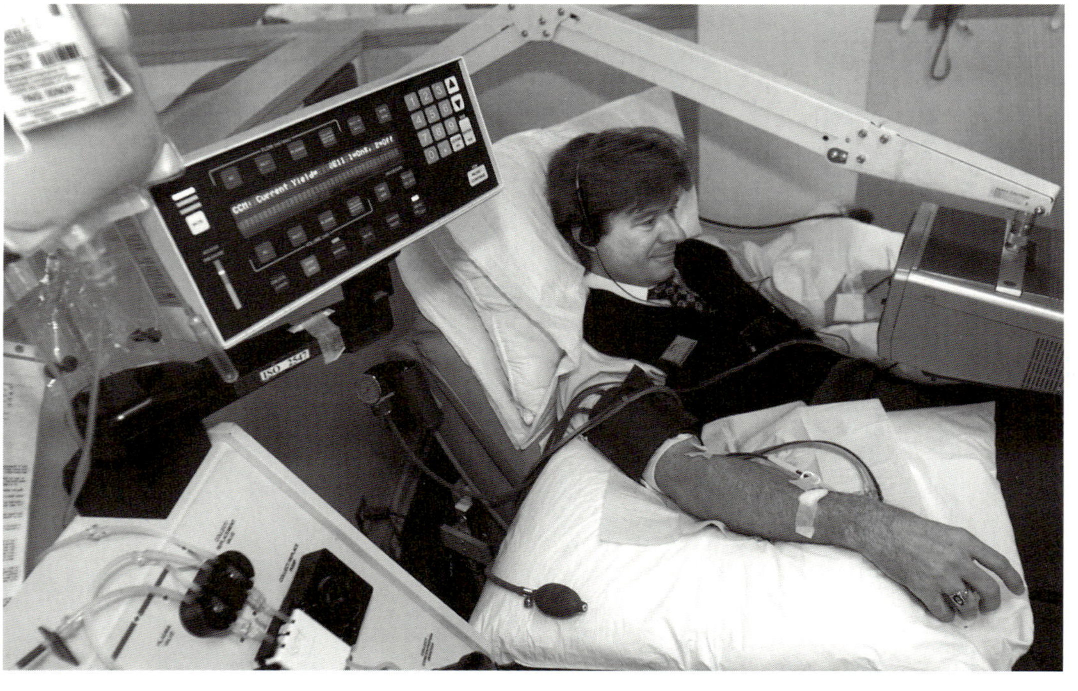

This donor is having platelets removed from his blood and stored for use by people with blood-clotting disorders. The machine removes the platelets and returns all the other blood components to the donor.

RED BLOOD TYPES

There are several different types of red blood cells. Every person in the world has just one particular type of red blood cell flowing through his or her body. The type of red blood inside a person's body can be recognized by certain antigens that are present or absent on the cell's surface. An *antigen* (AN-tih-jen) is a substance that causes the body's immune system to begin working. The practice of sorting red blood by types is known as *ABO blood-typing.* The most common blood types are A, B, AB, and O.

- A person with *type A* blood has cells with an antigen called marker A on their surfaces.
- A person with *type B* blood has cells with an antigen called marker B on their surfaces.
- A person with *type AB* blood has cells with both A and B markers on their surfaces.
- A person with *type O* blood cells has neither A nor B markers on their surfaces.

In addition, blood is grouped by whether or not it contains an antigen known as the *Rh factor.* A person with Rh in the blood is said to be *Rh positive,* while a person without Rh is said to be *Rh negative.* A person with type O blood who has Rh in the blood is said to be *O positive.* About 38 percent of all people in the United States have O positive blood, the most common blood type. The rarest type of blood is AB negative. Just 1 percent of all Americans have this type of blood.

Why is blood typing important? Every year, 4 million people receive blood transfusions. If a person were to receive blood that was not the correct type, he or she might have a reaction to the blood. The reaction could cause serious illness and death. People with type O negative blood are sometimes called "universal donors," because their donated blood can be used by people of any blood type. People with type AB positive blood are sometimes called "universal recipients," because they can receive any type of blood.

BLOOD PRESSURE

Blood pressure is a measurement that tells how much force, or pressure, blood is exerting against the walls of blood vessels. A person's blood pressure can vary, depending on the strength of the beating heart and where the pressure is being measured. This is because blood pressure is not the same in arteries, veins, and capillaries.

To be consistent, medical professionals measure the blood pressure of the main artery of the arm. They use an instrument called a *sphygmomanometer* (sfigg-moh-man-AH-meh-ter), as well as a stethoscope. A cuff is wrapped tightly around the arm, just above the elbow, and then pumped up with air. As the air is slowly released, doctors listen to the blood coursing through the artery.

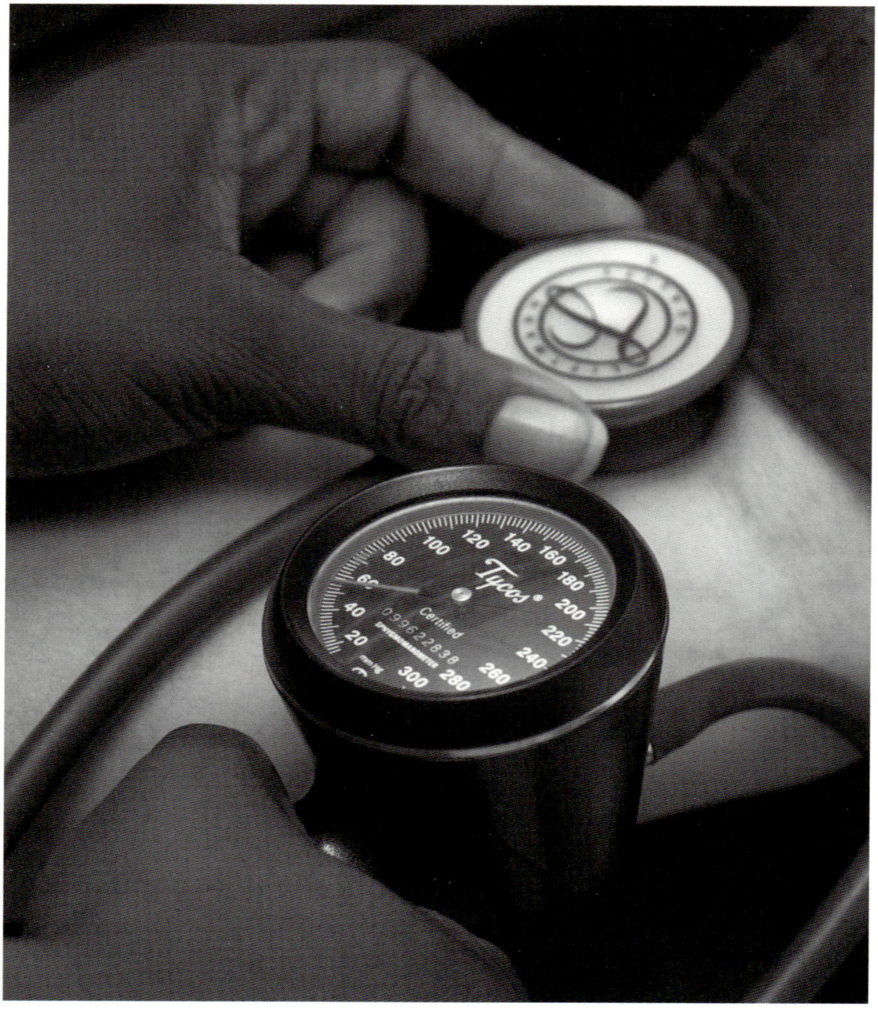

Blood pressure is routinely checked by health professionals using a sphygmomanometer like the one shown here.

Fascinating Circulatory Facts

- The average human body has about 62,000 miles (99,200 kilometers) of arteries, veins, and capillaries.
- It takes about one minute for blood to travel out of the heart, through the body, and back into the heart.
- When the body is resting, the heart pumps about seventy times each minute. However, this rate varies when the body is stimulated by stress, exercise, or adrenaline.
- Each second, 2 million red blood cells die and are replaced in the human body.
- The average male body contains from 5 to 7 quarts (4.7 to 6.6 liters) of blood. Women have a little less, from 4 to 6 quarts (3.8 to 5.7 liters).
- Blood travels at the rate of up to 10 miles (16 kilometers) per hour. That means that every hour, a blood cell travels 10 miles (16 kilometers) throughout the body.

For every blood pressure reading, the doctor takes two different measurements. *Systolic* (siss-TAHL-ik) *pressure* is the pressure on the artery walls when the heart is beating; *diastolic* (dye-uh-STAHL-ik) *pressure* is the pressure on the artery walls when the heart is between beats. Normal systolic blood pressure ranges from 100 to 120, while normal diastolic blood pressure measures between 60 and 80. Blood pressure is read as two numbers, with the systolic measure first and the diastolic second. For example, a person with normal blood pressure would have a reading of 120/80, or 120 over 80. Any person with a reading of 140/90 or higher is said to have high blood pressure, or hypertension. Doctors usually prescribe medication for people with a blood pressure reading of 145/90 or higher.

Fast Fact

There is no such thing as "blue blood." The blood in the veins only looks blue when it is seen through the skin.

Health and History: Understanding Circulation

Until the sixteenth century, most people had little understanding of how blood circulated throughout the body. In 1597, English physician William Harvey (1578–1657) set out to change that. He used *cadavers,* or dead human bodies, as well as live animals to learn about the heart and its role in moving blood through the body. In 1628, Harvey published his results. He became the first person to describe the heart as a pump. He also revealed that the blood flowed through the body in a circle, with different types of blood vessels carrying blood away from and to the heart.

2

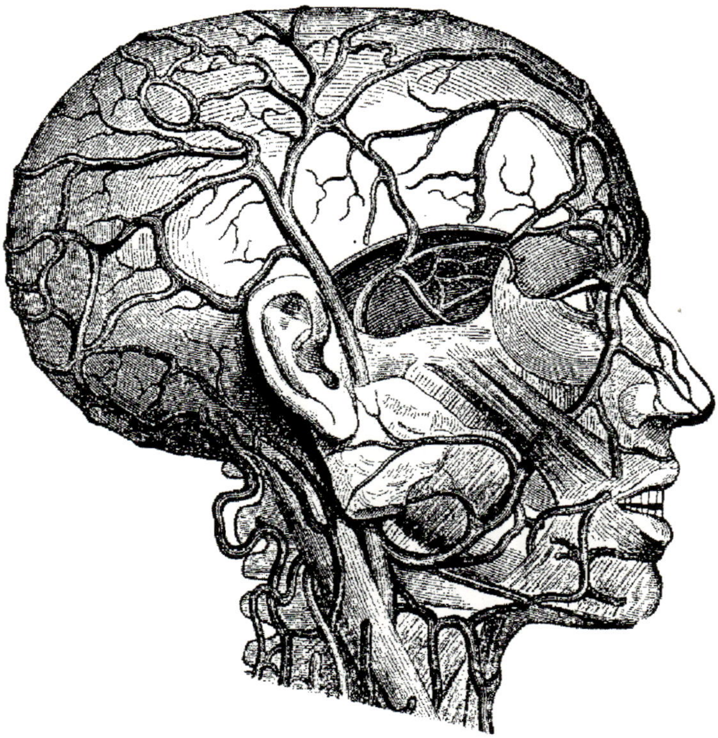

CIRCULATORY DISORDERS

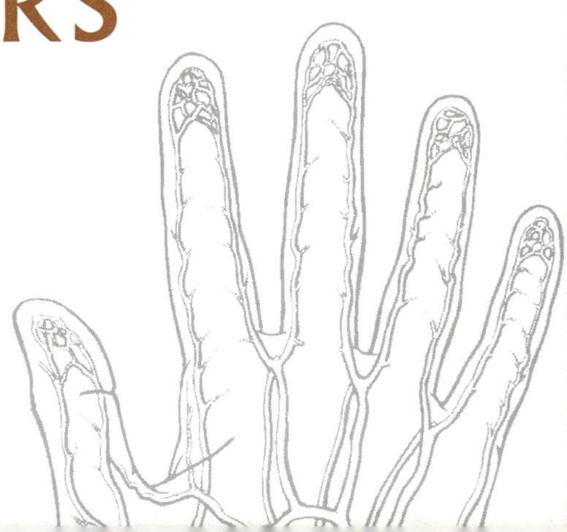

Medical problems can affect any part of the circulatory system: the heart, blood vessels, and blood. They may range from mild conditions that do not need treatment to life-threatening medical emergencies.

CONGENITAL HEART DISORDERS

Sometimes, people are born with heart problems. This occurs when the heart or the blood vessels near the heart don't develop properly before birth. These types of heart problems are called congenital heart disorders, or congenital heart diseases. The word *congenital* (kun-JEN-ih-tul) means "present at birth."

Congenital disorders can cause structural problems to the heart valves, heart muscles, or any other parts of the heart. Heart defects are the most common major *birth defects* among American babies. As many as one in a hundred babies is born with some type of congenital heart problem. Half are severe, requiring surgery within the first year of life. Some may cause potentially fatal conditions that must be attended to immediately. Other defects, however, are minor problems that lessen or disappear as a child grows.

Doctors are not sure what causes most congenital heart defects. Some may be caused by environmental factors. Exposure to certain chemicals or medications during pregnancy, for example, may lead to these birth defects. Other factors include exposure to bacterial or viral infections or the presence of other health conditions in the mother, such as diabetes or rubella. Birth defects may also be caused by a pregnant woman's lifestyle decisions, including smoking and drinking alcohol.

Symptoms of congenital heart defects include difficulty breathing, blue-tinged skin due to lack of oxygen in the blood, and heart failure. Children born with heart disorders may have to restrict certain types of physical activities. Some organized sports, for example, may be off limits. It is important for people suffering from heart defects always to check with a doctor before trying new physical activities.

TYPES OF CONGENITAL HEART DEFECTS

There are many different types of congenital heart defects, including the following:

NAME	Coarctation (koh-ark-TAY-shun)
NATURE OF DEFECT	Part of the aorta is narrowed or pinched, resulting in reduced blood flow.
SYMPTOMS	Dizziness, headaches, fainting, high blood pressure, nosebleeds, heart failure
TREATMENT	Surgery
NAME	Septal defects
NATURE OF DEFECT	Temporary holes between the two atria or ventricles fail to close properly before birth. Oxygenated blood leaks from the left side of the heart to the right side, and the flow of oxygen-rich blood throughout the body is reduced. The heart must work harder to compensate.
SYMPTOMS	Difficulty breathing, shortness of breath, frequent respiratory infections, failure to gain weight, fast heart rate
TREATMENT	Surgery
NAME	Patent ductus arteriosus (ar-teer-ee-OH-suss)
NATURE OF DEFECT	A temporary blood vessel near an infant's heart that is no longer needed after birth fails to close properly.
SYMPTOMS	Shortness of breath, poor eating habits, heart murmur, heart infections, heart failure
TREATMENT	Medicines to close the hole; surgery
NAME	Cyanotic (sye-uh-NAHT-ik) heart disease
NATURE OF DEFECT	Heart or blood vessel defects result in oxygen-poor blood circulating throughout the body.
SYMPTOMS	Cyanosis (sye-uh-NOH-siss), a bluish tinge to the skin that results from a lack of oxygen in the body
TREATMENT	Surgery
NAME	Transposition of the great vessels
NATURE OF DEFECT	The positions of the aorta and pulmonary artery are reversed. Blood circulating through the body does not carry enough oxygen to properly nourish tissues and organs.
SYMPTOMS	Difficulty breathing, cyanosis, poor eating habits, clubbed fingers or toes, heart murmur, problems with the heart valves
TREATMENT	Surgery

PROBLEMS WITH THE MYOCARDIUM

The *myocardium* (mye-oh-KAR-dee-um) is the muscle that makes up the walls of the heart's chambers. *Cardiomyopathies* (kar-dee-oh-mye-AHP-uh-theez) are diseases of this muscle. There are a number of different types of cardiomyopathy. Some are congenital, but they can also be caused by heart attacks, high blood pressure, dietary problems, and alcoholism. In some cases, the cause of such a condition is not known.

Congenital cardiomyopathies can cause the heart to become enlarged or weakened. These problems can affect one or both sides of the heart. Congenital cardiomyopathies interfere with the heart muscle's ability to pump blood effectively. Two such conditions include dilated cardiomyopathy and hypertrophic (hye-per-TRAH-fik) cardiomyopathy.

The symptoms of cardiomyopathy include shortness of breath, chest pain, fainting, and dizziness. A person suffering from this condition may also experience other symptoms, including fatigue and swelling of the legs, ankles, and abdomen. A more serious indicator of the disease is heart failure. Doctors use medications to treat cardiomyopathy. In some cases, surgery is necessary.

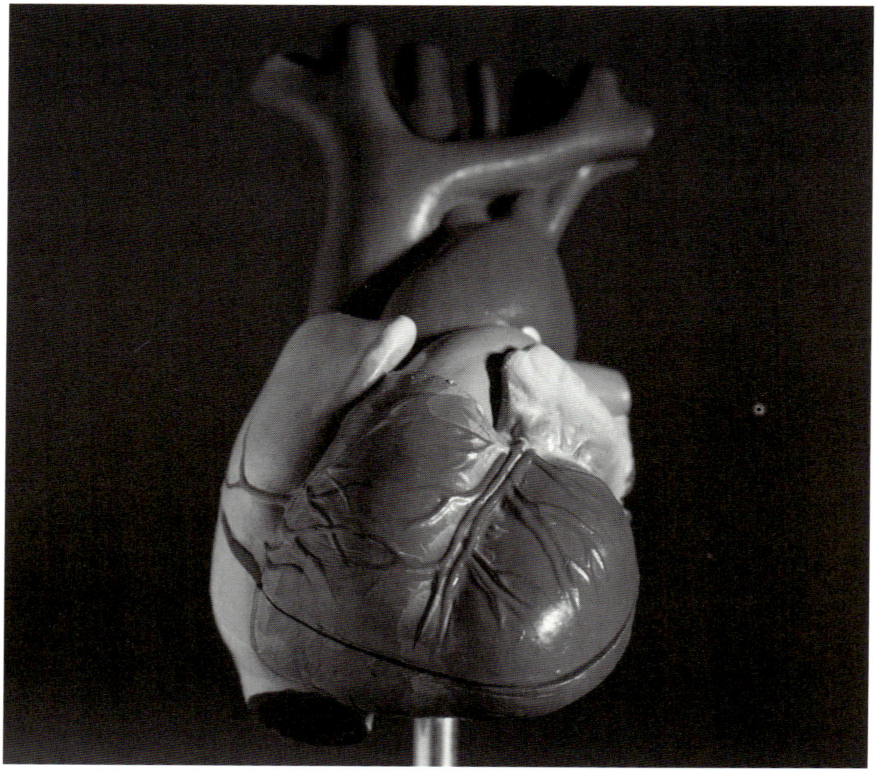

This model of the human heart is used to study the heart's structure and function. It shows the outside of the myocardium.

VALVE PROBLEMS

There are a number of birth defects that specifically affect the heart's four valves. These problems include the following:

NAME	Stenosis (sten-OH-siss)
NATURE OF DEFECT	A valve is too narrow, restricting blood flow.
SYMPTOMS	Some cases are asymptomatic (ay-simp-toh-MAT-ik), meaning there are no symptoms; when present, symptoms include chest pain, fatigue, dizziness
TREATMENT	Medication, and in serious cases, surgery
NAME	Regurgitation (ree-gur-jih-TAY-shun)
NATURE OF DEFECT	A valve does not close properly, allowing blood to flow back to where it has just come from.
SYMPTOMS	Many cases are asymptomatic; when present, symptoms may include high blood pressure; swelling of the abdomen, feet, and ankles; fatigue and weakness; weight loss; heart failure; cirrhosis
TREATMENT	Treatment unnecessary for asymptomatic cases; surgery for serious cases
NAME	Bicuspid (bye-KUSS-pid) aortic valve
NATURE OF DEFECT	The aortic valve has two cusps, or flaps that open and close to let blood in, instead of the normal three. As a result, the valve may not open or close completely.
SYMPTOMS	Many cases are asymptomatic; when present, symptoms may include increased valve infections, weakness, fatigue
TREATMENT	Surgery later in life, if the condition causes serious problems

One of the first signs of a congenital heart defect may be a *heart murmur.* Heart murmurs are vibrations that result from turbulent blood flow in the heart. This chaotic blood flow makes a whooshing or rasping sound that can be heard through a stethoscope. A heart murmur may be caused by a defective valve or other problem. Some heart murmurs are warning signs of a serious medical condition. Others are harmless or "innocent."

Since the mid-1960s, doctors have been using artificial heart valves to replace defective valves and save lives. Artificial heart valves are made out of a variety of substances, including metal, plastic, and even human or animal tissue. Those placed in the body are generally expected to function from ten to twenty years. People with artificial heart valves run a higher than normal risk of blood clots and infections. To ward off these problems, they must take medicines that help them remain healthy.

Health and History: Highlights in Heart Surgery

Techniques of heart surgery improved rapidly during the twentieth century. Here are a few of the most important moments.

- 1938: The first successful heart surgery is performed by Robert Gross (1905–1988) in Boston, Massachusetts. Gross corrects a patent ductus arteriosus in a seven-year-old girl.
- 1952: The first artificial heart valve is implanted into a human by Charles Hufnagel (1916–1989) in Washington, D.C. The valve is made of plastic.
- 1952: F. John Lewis (1916–) and Walton Lillehie (1918–1999) perform the first successful open-heart surgery on a five-year-old girl at the University of Minnesota.
- 1953: The heart-lung machine, developed by John Gibbon (1903–1974) is used for the first time in Philadelphia, Pennsylvania, to keep a patient alive during heart surgery.
- 1967: The first successful heart transplant is performed by Christiaan Barnard (1922–2001) in Capetown, South Africa. Unfortunately, the patient lives for just eighteen days after the surgery.
- 1982: The first permanent artificial heart is transplanted into a patient by William DeVries (1943–) at the University of Utah Medical Center. The patient lives for 112 days with a Jarvik-7, invented by Robert Jarvik (1946–) of Salt Lake City, Utah.

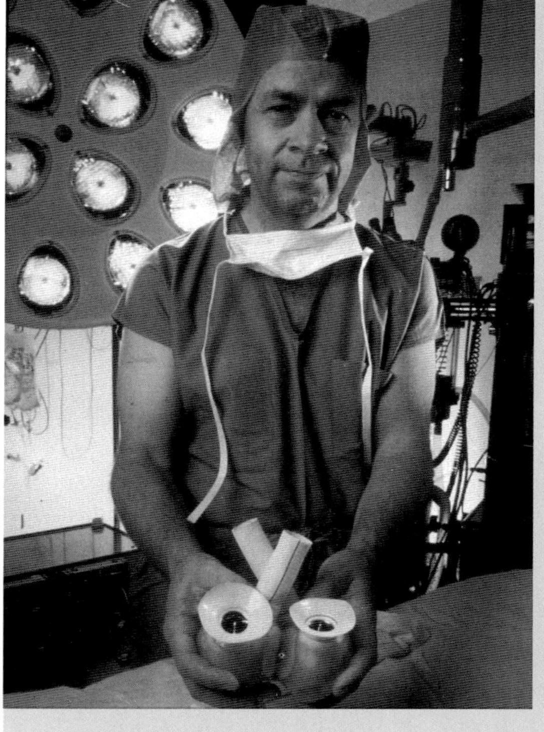

Dr. Jack C. Copeland holds a Jarvik-7 artificial heart in the operating room of the University of Arizona Medical Center in Tucson.

CONGENITAL BLOOD AND BLEEDING DISORDERS

Some people are born with disorders that affect other parts of the circulatory system besides the heart. Blood disorders include various problems with a person's blood or the production of blood. Bleeding disorders are conditions in which the blood does not clot properly and continuous bleeding occurs. Bleeding disorders can lead to damage and scarring of the joints, vision problems, neurological problems, and chronic anemia. *Anemia* (uh-NEE-mee-uh) is a lack of red blood cells in the body.

HEMOPHILIA

Hemophilia (hee-moh-FILL-ee-uh) is one of several inherited bleeding disorders. With this condition, the body doesn't produce proteins that help blood clot. As a result, the blood clots very slowly, and blood from a wound may continue to flow for a long time.

Males are much more likely to have hemophilia than females. However, females carry the genetic defect that causes this condition. It is possible for a female to get hemophilia, but only if her mother and father both possess the defective gene that causes the disease.

Symptoms of hemophilia include raised bruises, prolonged nosebleeds, excessive bleeding, and blood in the urine. Bleeding frequently occurs inside the body, in joints and muscles. This can cause discomfort, aching, and pain. It can also lead to permanent musculoskeletal (mus-kyoo-loh-SKEL-uh-tul) damage.

Doctors diagnose hemophilia after conducting a *complete blood count* (CBC). By learning which proteins are missing, they can determine whether a patient has hemophilia and what type it is. Would-be parents who know that they have the genetic abnormality that causes hemophilia may choose *prenatal* screening. Tests are conducted on the fetus in the womb to determine if the baby has hemophilia.

In the past, children with hemophilia were forced to avoid physical activity for fear of seriously harming themselves. Today, doctors treat people with hemophilia by giving them transfusions of the missing clotting factor. Medicines are also available to help the body produce the missing protein.

Health and History: The Royal Disease

Experts think that the first mention of the disease hemophilia appeared in writing nearly 2,000 years ago. Since that time, physicians have worked to learn more about this disease. In the nineteenth century, interest in hemophilia was raised when Queen Victoria (1819–1901) was diagnosed as having it. Victoria passed the abnormal gene for this disorder down to her children. One son, Leopold (1853–1884), died as a result of hemophilia. Victoria's other children married into royal families throughout Europe, spreading the disease. Hemophilia soon earned the nickname "the Royal Disease."

VON WILLEBRAND DISEASE

Von Willebrand disease is the most common type of bleeding disorder. Unlike hemophilia, both men and women are affected nearly equally by von Willebrand disease. As many as 2 percent of the population is affected by this condition.

Von Willebrand disease occurs when platelets do not stick properly to damaged blood vessels. Because this important first step in clotting is missing, affected people do not heal as quickly as they should. Doctors recognize three forms of von Willebrand disease that vary in severity.

Symptoms of von Willebrand disease are similar to those of hemophilia. They include frequent bruising and excessive bleeding from wounds. Women may experience excessive bleeding during their menstrual periods.

People with von Willebrand disease must take special precautions before having any type of surgery. They must take special medicines beforehand to help their blood clot so the surgical incision can heal. Some patients may need to receive infusions of plasma and clotting proteins to help them recover.

SICKLE-CELL ANEMIA

Sickle-cell anemia, or sickle-cell disease, is an inherited type of anemia in which the red blood cells become abnormally shaped. Instead of being round, they are long and sickle-shaped. Sickle-cell anemia is caused by an abnormal type of hemoglobin.

The abnormal red blood cells have difficulty passing through the blood vessels. As a result, people with sickle-cell anemia are likely to suffer frequent and painful blood clots. Other symptoms of sickle-cell disease include pain, fever, swollen joints, jaundice (yellowness of the skin), and eye problems. In severe cases, sickle-cell anemia can lead to death as a result of infection and organ failure.

Sickle-cell anemia is a *genetic disorder* that can only occur when both parents have the genetic abnormality that causes the disease. It most commonly affects African Americans. It is estimated that one in every eight African Americans has the trait that causes sickle-cell anemia, and one in every 500 African American infants is born with the disease. Health officials encourage people who believe they may be at risk for sickle-cell anemia to be tested before having children.

Doctors test for this disease by drawing blood and checking for the abnormal red blood cells. Treatment may include medicine to ease aches and pains, antibiotics to fight off infections, blood transfusions, and bone-marrow transplants.

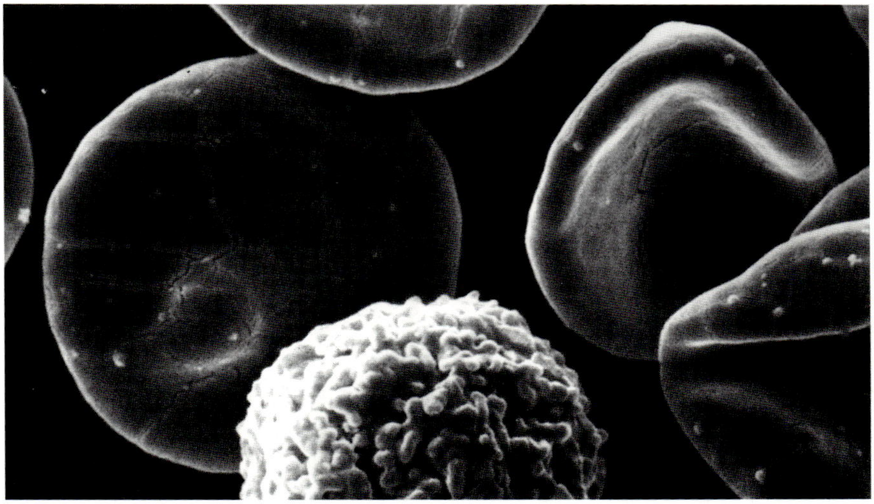

This microscopic photograph shows a cell afflicted with sickle-cell anemia alongside other healthy cells.

THALASSEMIA

Thalassemia (thal-uh-SEE-mee-uh) is a type of anemia that most often affects people of Asian and Mediterranean descent. Like sickle-cell anemia, thalassemia is marked by abnormally shaped red blood cells. The abnormal cells are prematurely destroyed by the body. Symptoms of the condition include shortness of breath, fatigue, jaundice, and deformations of the face. Severe cases can lead to iron buildup in the body, which may damage the liver, heart, and endocrine system. Death may result from heart failure due to such buildup. Doctors treat the condition by giving patients blood transfusions and folic-acid supplements.

OTHER CIRCULATORY CONDITIONS

Not all circulatory problems are congenital. Some are caused by disease, other health conditions, or aging. Others are *autoimmune* (aw-toh-im-MYOON) *disorders,* which occur when the body's immune system mistakenly attacks healthy tissue in the body. The causes of some conditions remain unknown. Here are a few circulatory diseases that are not genetic.

INFLAMMATION OF THE BLOOD VESSELS

When blood vessels become inflamed and swollen, normal blood flow through the body is decreased. Inflammation of the blood vessels, known as *vasculitis* (vass-kyoo-LYE-tiss), may be caused by infection, injury, or other conditions. Some types of vasculitis are believed to be autoimmune disorders. When the immune system mistakenly identifies something in a blood vessel as a foreign body, it attacks the substance, damaging the blood vessel in the process. Triggers that set off this reaction include some illegal drugs, such as cocaine and amphetamines. Infections may also trigger an abnormal immune system response.

The symptoms of vasculitis vary, depending upon which blood vessels and organs are affected. Vasculitis of the blood vessels in the brain, for example, may cause paralysis, vision problems, and strokes. *Paralysis* (puh-RAL-ih-siss) is the loss of movement or sensation in part of the body. A *stroke* occurs when the blood supply to the brain is cut off, resulting in tissue damage or even tissue death. Doctors treat vasculitis with steroids and other medications. Any affected or damaged organs must also be treated.

To diagnose vasculitis, doctors use blood tests. They look for *antibodies,* substances that signal an immune system reaction, in the blood. Angiography, magnetic resonance imaging (MRI), and X-rays may also be used. Types of vasculitis include the following:

NAME	Kawasaki (kah-wah-SAH-kee) disease
NATURE OF DEFECT	The lining of the blood vessels, lymph nodes, heart, and mucous membranes (layers of mucus-secreting cells that line the tubes and cavities of the body) are affected.
SYMPTOMS	Fever, rashes, swollen hands and feet, swollen lymph nodes, irritation of the mouth, lips, and throat, heart damage
TREATMENT	Hospitalization and intravenous medications
NAME	Behcet's (buh-SETZ) disease
NATURE OF DEFECT	The mouth, eyes, and genitals are affected.
SYMPTOMS	Ulcers (sores) of the mouth, skin lesions, eye infections, stroke, memory loss, blood clots, arthritis (arth-RYE-tiss)
TREATMENT	Medication to treat symptoms; there is no cure.
NAME	Polyarteritis nodosa (pahl-ee-ar-tuh-RYE-tiss noh-DOH-suh)
NATURE OF DEFECT	Small and medium-sized arteries become inflamed, resulting in tissue and organ damage.
SYMPTOMS	Heart, skin, kidney, and nervous system damage may lead to fever, fatigue, weight loss, muscle and joint aches, rashes, numbness, seizures, stroke, kidney failure, heart failure, death
TREATMENT	Medication to treat symptoms; there is no cure.
NAME	Wegener's granulomatosis (gran-yoo-loh-muh-TOH-siss)
NATURE OF DEFECT	Blood vessels in the respiratory tract swell. The lungs, kidneys, musculoskeletal system, and skin are also affected.
SYMPTOMS	Sinus pain, ulcers inside the nose, arthritis
TREATMENT	Medication to suppress the immune system.
NAME	Henoch-Schonlein purpura (HEH-nahk SHONE-lyne PUR-pur-uh)
NATURE OF DEFECT	The capillaries become inflamed, causing bleeding into the skin, lungs, or intestines.
SYMPTOMS	Rashes, aching joints, stomach pain, fever, vomiting, swelling of the ankles or scrotum, blood in the urine
TREATMENT	Medication
NAME	Giant cell arteritis (ar-tuh-RYE-tiss)
NATURE OF DEFECT	The arteries in the upper body (especially the temple, head, neck, and arms) become inflamed as a result of a buildup of abnormally large (giant) cells
SYMPTOMS	Headaches, fever, pain or tenderness in the affected area, vision problems, pain while chewing, blindness, strokes
TREATMENT	Medication

ARTERIOSCLEROSIS

Arteriosclerosis (ar-teer-ee-oh-skluh-ROH-siss) is a disease that affects the arteries and sometimes the veins. More commonly known as "hardening of the arteries," this condition causes blood vessels to become thicker and less flexible. Arteriosclerosis is a common and serious health problem. It is the leading cause of heart disease in the United States today.

According to the Vascular Disease Foundation, arteriosclerosis is a natural result of aging. When the human body is young, the arteries are clear and elastic. As the body ages, however, the arteries become thicker and stiffer. Types of arteriosclerosis include atherosclerosis, cerebral arteriosclerosis, and carotid artery disease.

Atherosclerosis (ath-er-oh-skluh-ROH-siss) is caused by a buildup of plaque in the blood vessels. *Plaque* is a sticky substance made up of fats and cholesterol, as well as blood cells and other body tissues. As plaque builds up, blood vessels weaken and clog, preventing normal blood flow to the various parts of the body. Unfortunately, many people don't realize that they have atherosclerosis until it causes a major medical crisis, such as heart attack, heart failure, or stroke.

Peripheral (pur-IFF-er-ul) *vascular disease* (PVD) is also known as peripheral arterial disease (PAD). This condition is a type of atherosclerosis that affects the *extremities* of the body—the arms, legs, hands, and feet. Arteries that carry blood to these areas become thin and hard. As a result, there is an increased risk of blood clots in the narrowed arteries. Symptoms of PVD include leg cramps, pain, numbness, tingling, skin ulcers, and the loss of hair on the legs and arms.

PVD is very common in the United States. About 8 million people suffer from this condition. As with many other arterial diseases, people who smoke, are obese, eat unhealthy diets, or don't exercise are at risk for this condition. Diabetes and hypertension can also increase the risk.

Cerebral (suh-REE-brul) *arteriosclerosis* affects the arteries in the brain. This condition can cause strokes; aneurysms, or abnormal swelling of blood vessel walls; and *hemorrhaging* (HEM-or-uh-jing), or bleeding, in the brain. Symptoms of cerebral arteriosclerosis include headache, pain in the facial area, and vision problems. To treat this type of arteriosclerosis, doctors may use drugs or surgery. Doctors also recommend lifestyle changes, including quitting smoking and eating a healthy diet.

The carotid artery carries blood and oxygen from the heart to the brain. With carotid artery disease, plaque builds up in this blood vessel, and a stroke can result. Surgery is often required to correct this problem before serious damage is done.

VARICOSE VEINS

Varicose veins result when valves in the veins of the lower legs become defective. This results in blood draining backward and collecting in the superficial veins near the skin's surface. Varicose veins can be quite painful, causing swollen feet and ulcers of the skin. In some cases, surgery is required to ease the condition. Varicose veins can lead to other conditions, including phlebitis, blood clots, ulcers, and rupturing of the veins.

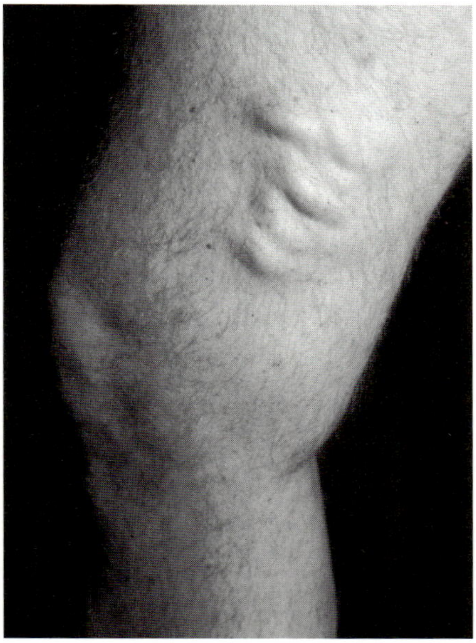

Varicose veins can be seen protruding from this person's leg.

PHLEBITIS

Phlebitis (fleh-BYE-tiss), also known as thrombophlebitis (thrahm-boh-fleh-BYE-tiss), is a swelling in one or more veins. The inflammation is often caused by a blood clot. Because phlebitis is frequently linked to varicose veins, clots that cause the inflammation usually occur in the legs. Phlebitis may also be caused by bacterial infection, prolonged inactivity, injury, or an unknown cause. Symptoms of phlebitis include pain, warmth, and tenderness in the affected area. Legs may become red and swollen if the case is severe.

To treat the condition, doctors work to reduce the inflammation. They may use painkillers and medicines that thin the blood and help break up blood clots. Other methods of treatment include warm compresses and elevation of the leg. In serious cases, surgery to remove or bypass a swollen vein may be necessary.

GANGRENE

Gangrene occurs when circulation to a section of the body is cut off. When this happens, tissue and cells in that part of the body die and begin to decay. Frostbite, infections, injuries, and diabetes can all cut off circulation and cause gangrene.

Gangrene is a serious medical condition. Once tissues have died, they cannot be repaired. Doctors can treat some cases of gangrene with medicines. In other cases, doctors may recommend surgery to remove the dead tissue. Dead tissue must be removed to prevent the spread of infection to other parts of the body. In the most severe cases of gangrene, amputation is necessary. *Amputation* is the removal of a dead or diseased limb, especially by surgery.

Two types of gangrene involve problems with the circulatory system. *Wet gangrene* is caused by a sudden interruption of blood flow to a part of the body. It most often occurs as a result of severe heat or cold, injury, or a serious blood clot. The affected areas appear bruised, swollen, or blistered. If infection sets in and spreads, wet gangrene can be fatal. *Dry gangrene* is caused by a reduced flow of blood through the arteries of the arms or legs. This type of gangrene occurs gradually and is signaled by dry, dark skin near the affected area. Dry gangrene is usually not life threatening.

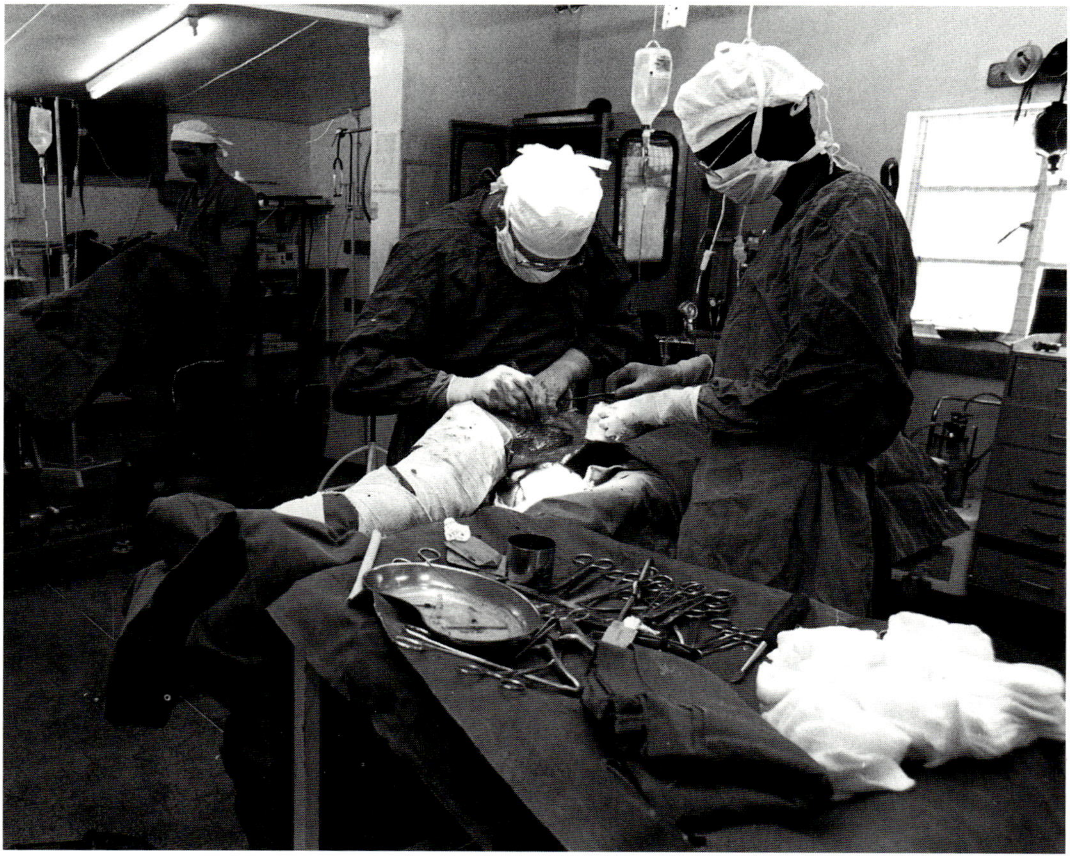

Surgeons amputate the gangrenous leg of a land mine victim.

DECOMPRESSION SICKNESS

Also known as the bends or caisson disease, *decompression sickness* most often affects deep-sea divers. As divers descend deeper and deeper into the ocean, greater amounts of nitrogen dissolve in the blood. When the divers return to the surface, they must rise slowly, even stopping several times along the way. This slow, gradual ascent allows the nitrogen to move out of the bloodstream and be exhaled through the lungs. When a diver swims upward too quickly, nitrogen is released from the bloodstream as gas bubbles. These gas bubbles can injure tissue and block blood vessels.

Symptoms of decompression sickness include shaking, nausea, and pain and stiffness in the affected joints. When bubbles form in the brain, a diver may become paralyzed. Doctors treat those suffering from decompression sickness by placing the patient in a special, pressurized chamber.

EMBOLISMS

Embolisms (EM-boh-liz-umz) occur when particles in the bloodstream become stuck in a blood vessel. These particles could be clots or parts of clots that have broken free. They might also be cholesterol crystals, fat from fractured bones, or any other pieces of matter. The blockage, called an *embolus,* must be removed.

One particular type of blockage is a *pulmonary embolism.* A pulmonary embolism occurs when a clot travels from its original site, passes through the heart, and lodges in a pulmonary artery of the lungs. This causes a lack of blood flow and oxygen to the lungs. Most clots start in the legs, but they can come from other sections of the body, as well.

A pulmonary embolism is a serious medical condition that can cause death. Each year, about 600,000 Americans suffer pulmonary embolisms, and 60,000 die. Symptoms of the condition include difficulty breathing, chest pain, rapid pulse, sweating, and fainting. It is important for people with these symptoms to seek medical treatment immediately.

People at high risk for pulmonary embolisms include those who have been confined to bed or who have been inactive for a long period of time. People on long-distance plane trips, for example, have been known to suffer from this condition. Other risk factors include surgery, childbirth, the use of birth control pills, and many health conditions. To prevent embolisms, doctors recommend a low-fat, high-fiber diet and plenty of exercise.

ANEURYSMS

An *aneurysm* (AN-yoor-iz-um) is a swelling of the wall of an artery. As the wall swells, it creates a balloon or bubble in the artery. The swelling grows larger and larger, until the balloon eventually bursts, causing stroke or even death. Because there are no symptoms associated with this condition, most aneurysms are detected only during a regular physical examination.

Three out of four aneurysms are *abdominal aortic aneurysms* (AAA). This type of arterial condition occurs when an aneurysm develops along the part of the aorta that passes through the abdomen. When an AAA bursts, the chances of death as a result are very high. Other common aneurysm sites include the brain, chest, and legs. When a brain aneurysm bursts, it is called a *cerebral hemorrhage.* Sometimes, when pieces of plaque break away from inside the balloon and become lodged in the blood vessels of the legs, aneurysms can lead to gangrene.

The exact cause of most aneurysms is unknown. Some may be caused by genetic abnormalities. Others may be caused—or worsened—by high blood pressure. In addition, recent research has shown a strong link between ruptured aneurysms and smoking.

To prevent a ruptured aneurysm, surgery before the swelling bursts is desirable. Surgery to repair an aneurysm includes opening the aneurysm and removing its contents. Then a patch or a flexible, artificial tube, called a *graft,* is placed into the body to replace and strengthen the blood vessel. The artery wall is closed and stitched together.

DIAGNOSING CIRCULATORY PROBLEMS

Today, doctors have a number of different tools at their disposal to help them diagnose various circulatory problems. For example, blood tests can be used to check for evidence of heart attacks or other conditions that may be causing heart problems. Doctors can also take chest *X-rays.* X-ray machines use electromagnetic *radiation* (energy in the form of rays or particles) to create a picture of the heart. Doctors can examine the size and shape of the heart and its connection to the lungs. They can also use chest X-rays to see any *congestion,* or fluid buildup, in the lungs.

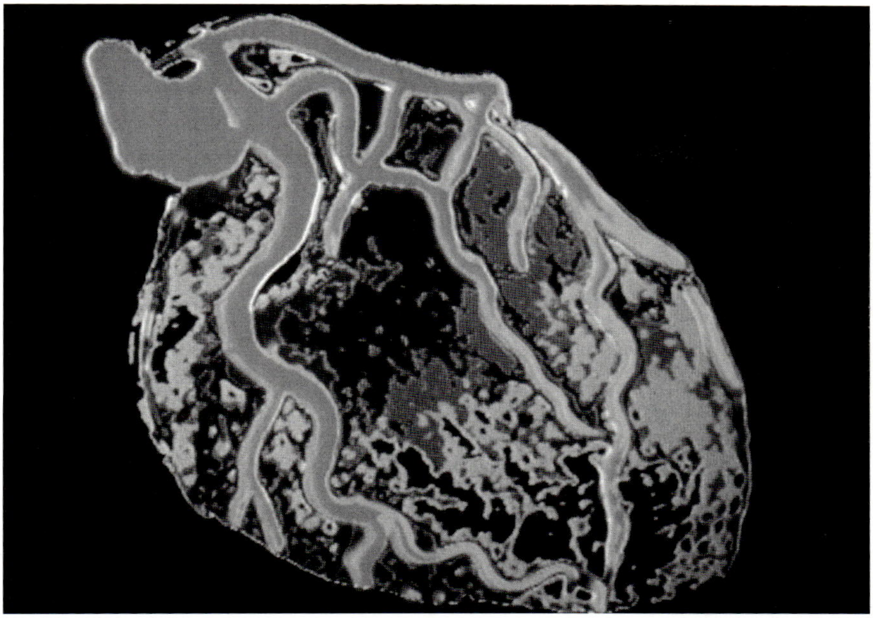

High-frequency sound waves in this echocardiogram show the structures of the heart.

Doctors can also perform an *echocardiogram* (ek-oh-KAR-dee-oh-gram). This test uses high-frequency sound waves to create a picture of the heart. Doctors can use the test to check the size, shape, and motion of the heart. Measurements may be taken while a person is resting or exercising.

In an electrocardiogram (ee-lek-troh-KAR-dee-oh-gram), or ECG, electrodes attached to the skin of the chest measure the electrical activity of the heart. Some people wear the electrodes and a portable monitor for twenty-four hours, allowing doctors to check heart activity during all sorts of routine activities. Electrocardiograms can detect heart murmurs and provide other information about the health of the heart. However, this type of test cannot reveal the cause of a heart problem.

Doctors can also use stress tests to put a patient's heart in a "stressed" condition so they can see how well the circulatory system is working. In some cases, the heart is stressed by having the patient exercise on a treadmill or a stationary bicycle. Those who cannot exercise are given medications that duplicate the effect of exercising on the heart.

Coronary angiography (COR-uh-ner-ee an-jee-AH-gruh-fee), also known as arteriography (ar-teer-ee-AH-gruh-fee), involves inserting a *catheter*, or small tube, into the heart or an artery (often in the groin) that feeds into the heart. Dye is injected into the catheter, and an X-ray machine captures pictures of the heart and blood vessels. The dye allows doctors to check the heart's structure and functioning. With this test, doctors can look at how the blood is circulating and detect blockages in the blood vessels.

MRI angiography (MRA) uses a *magnetic resonance imaging* (MRI) machine to look into the body and examine the blood vessels. An MRI machine uses magnets and radio waves to create an image of the inside of the body. Unfortunately, an MRI is more expensive than an X-ray, and many people with little or no health insurance can't afford it.

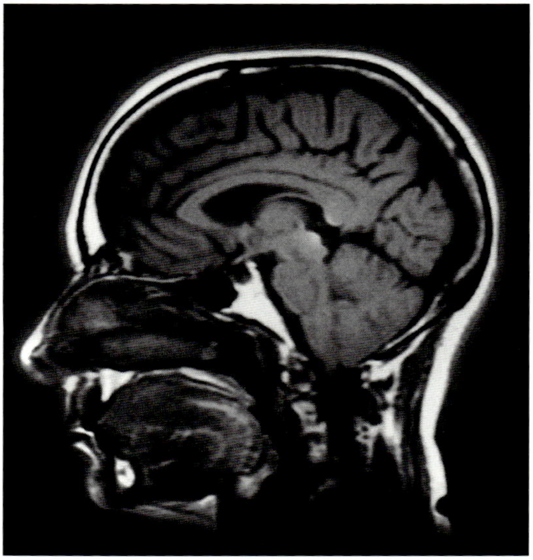

Magnetic resonance imaging (MRI) reveals the soft tissue structures of this woman's head, including her brain. Doctors use these images to look for abnormalities without the use of surgery.

In a *nuclear heart scan,* radioactive particles are either injected into a vein or inhaled into the lungs. Then a scanner or camera that can detect the radioactive particles is used to search for a pulmonary embolism and other heart problems.

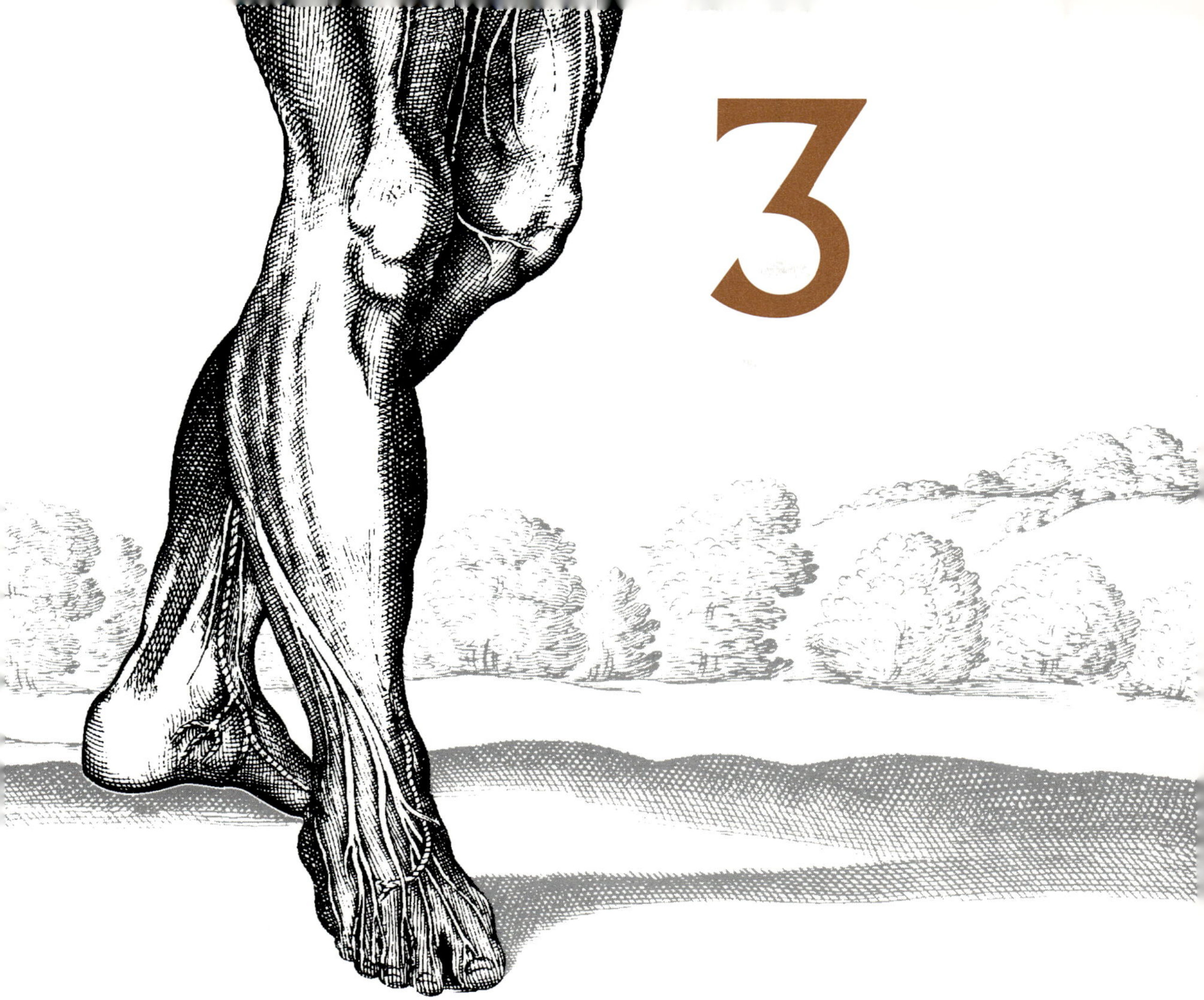

3

HOW THE ENVIRONMENT AFFECTS THE CIRCULATORY SYSTEM

The world around us can affect the health of the circulatory system and the rest of the human body. Environmental factors that can contribute to disease and other health conditions include pollutants; chemicals; diet and lifestyle choices; and bacteria, viruses, and other microorganisms.

CHEMICALS AND DISEASE

Exposure to chemicals and other toxins can cause some serious circulatory problems. Many of these pollutants enter the bloodstream after being inhaled through the lungs. Others enter through pores and wounds in the skin. No matter how they get into the circulatory system, many can cause diseases and other health problems. The following are a few circulatory diseases that can be caused by chemicals and other toxins in the body.

RAYNAUD'S PHENOMENON

Raynaud's phenomenon (ray-NOZE fuh-NAH-meh-nahn) is a condition that affects the small blood vessels. The vessels periodically spasm and become narrow, cutting off the flow of blood and oxygen to the extremities. People suffering from Raynaud's phenomenon may experience tingling and whiteness in the fingers, ears, nose, and lips. The skin in these areas may then turn blue. As the blood begins circulating again, the extremities turn red and become painful.

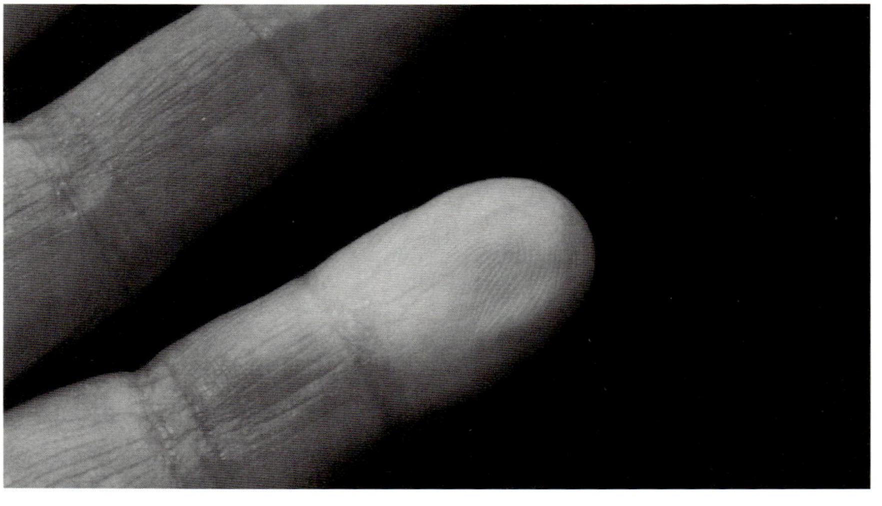

White fingertips caused by restriction of the blood vessels are a sign of Raynaud's phenomenon.

Although the exact cause of this condition is unknown, people who are exposed to certain drugs and chemicals are at increased risk. These substances include beta blockers, some migraine medications, some chemotherapy drugs, and vinyl chlorides. Other high-risk factors include smoking, typing, and other job-related activities. In addition, Raynaud's can be caused by cold and stress. This condition may occur by itself, or it may occur along with conditions of the musculoskeletal system, including lupus, rheumatoid (ROO-muh-toyd) arthritis, carpal tunnel syndrome, or scleroderma (sklur-uh-DUR-muh). In some cases, medication is necessary. In severe and debilitating cases, surgery may be necessary.

LEUKEMIA

Leukemia (loo-KEE-mee-uh) is cancer of the blood and blood-forming tissues. When a person has leukemia, the bone marrow produces too many white blood cells, and many are abnormal. It is estimated that people with leukemia produce fifty to sixty times more white blood cells than normal.

As the body produces more and more white blood cells, there is room for fewer red ones. This results in anemia, or a lack of red blood cells. The white blood cells begin to block out red blood cells, plugging up blood vessels that feed important organs. Because the white blood cells are abnormal, they do not correctly perform their task of protecting the body.

Some types of leukemia affect children between two and eight years old. Each year, about 2,200 children are diagnosed with this disease. However, most people affected by the condition are adults. Every year, 27,000 American adults learn that they have leukemia. Depending upon the type of leukemia, it may spread slowly (chronic) or quickly (acute).

Early symptoms of leukemia include fatigue, difficulty breathing, swollen lymph nodes, excessive bruising or bleeding when injured, fever, and frequent infections. People who notice early symptoms of the disease may think that they have the flu. The cancer may eventually spread to other parts of the body, including the brain, lymph nodes, liver, and spleen. Without successful treatment, leukemia can be life threatening.

The exact cause of leukemia is unknown. Some cases of leukemia occur as a result of a random genetic abnormality. Doctors have identified a number of risk factors that increase the chances of getting this type of cancer. For example, children with certain other genetic conditions, including Down syndrome, are more at risk of developing leukemia than other children are.

Exposure to certain environmental agents can also place a person at increased risk of leukemia. Intense radiation, for example, can act as a trigger for leukemia. As a result, people who have received radiation treatment for other cancers are at a greater risk. Exposure to some chemicals, such as mustard gas and benzene, also increases the risk of getting leukemia. Doctors believe that smoking can be linked to an increased risk of this cancer, as well.

DIAGNOSING LEUKEMIA

Symptoms of leukemia in children include a larger than normal number of fevers, infections, bruises, and nosebleeds. Patients may also experience fatigue, aching joints or bones, and swollen lymph nodes. Once leukemia has spread to other parts of the body, more serious symptoms may occur. For example, if the leukemia has spread to the brain, a patient may experience headaches, seizures, and problems with vision and balance.

Doctors determine whether a person has leukemia by using one or more tests. A doctor who suspects that a patient has leukemia will ask for a CBC. The blood is examined for the number of white and red blood cells. The blood will also be examined for the abnormal white cells that occur with leukemia.

To determine what type of leukemia a person is suffering from, doctors may perform a *bone marrow aspiration.* For this test, doctors use a long needle to extract a small amount of liquid marrow from a long bone in the body. The bone marrow is then examined under a microscope for any abnormalities that may signal leukemia. For a *bone marrow biopsy* (BYE-ahp-see), doctors use a larger needle to extract a bit of bone in addition to bone marrow. Both the bone and the bone marrow are examined under the microscope for abnormalities. For a *lymph node biopsy,* lymph nodes are removed from the body to be checked for abnormal cells.

During a *lumbar puncture* (spinal tap), spinal fluid is removed from the lower back and examined. The presence of abnormal cells in the spinal fluid indicates that the leukemia has spread to the brain and central nervous system.

TREATING LEUKEMIA

Leukemia, although not preventable, is curable in many cases. Some types of leukemia have *remission* rates of up to 90 percent. The remission rate represents how many people temporarily recover from the cancer.

Doctors treat leukemia with *chemotherapy* (kee-moh-THERR-uh-pee). Chemotherapy is the use of cancer-fighting drugs to stop the cancer. However, long-term chemotherapy treatment can cause serious health problems. It damages the bone marrow and some cells, resulting in anemia and a higher risk of infections. Patients undergoing chemotherapy often lose their hair and may experience severe nausea and vomiting. Long-term effects of chemotherapy may include damage to the heart or kidneys.

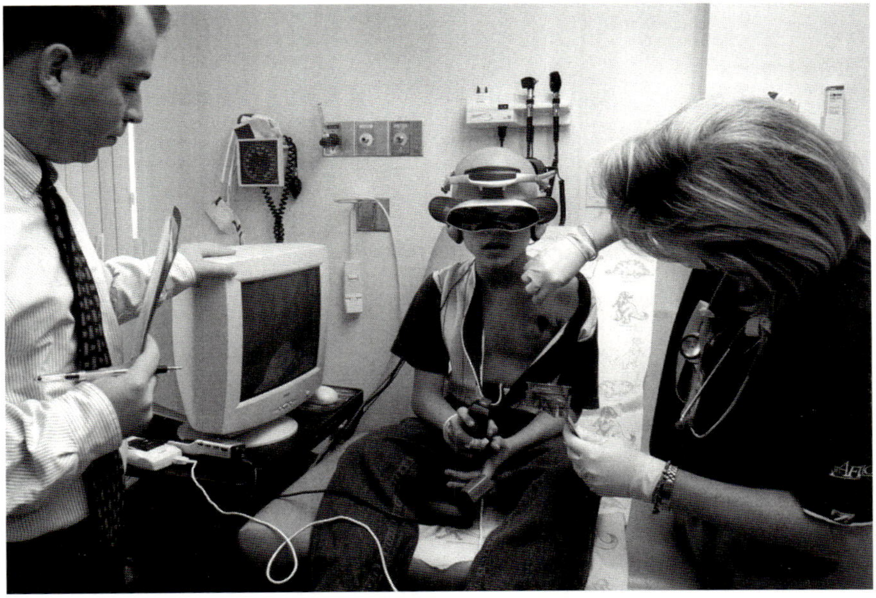

This young patient is being prepared for chemotherapy treatment. He is wearing a virtual reality helmet that is not part of the treatment. The virtual reality game helps to distract him from the adverse physical effects of the treatment.

Other courses of treatment in addition to chemotherapy may include radiation to shrink tumors. Some cases of leukemia may also require bone-marrow transplants. The goal of this procedure is to replace the patient's leukemia-producing bone marrow with healthy marrow. First, doctors must use drugs and radiation to destroy the patient's bone marrow. Then healthy new marrow taken from a donor is injected into the bone.

Some doctors choose to perform stem-cell transplants after chemotherapy. *Stem cells* are cells in the body that produce healthy new cells and tissue. In this procedure, doctors remove healthy stem cells from a person who is about to begin chemotherapy. After chemotherapy, the stem cells are injected back into the body, spurring new cell growth. This same type of procedure can also be performed with antibodies that fight leukemia.

Stem-Cell Research

Stem cells are "blank cells" that eventually develop into all of the different types of cells and tissues in the human body. Some of the most powerful stem cells are *embryonic* (em-bree-AHN-ik) *stem cells.* These cells are found in the bodies of developing embryos and fetuses.

One way that embryonic stem cells are obtained is from aborted fetuses. Many people object to the use of these stem cells. These people feel that the embryos represent human lives and that destroying a human life for any reason is wrong. They say that using these cells is immoral.

Another source of stem cells is from embryos that are created in a laboratory for in vitro fertilization. *In vitro fertilization* is a process in which an embryo is implanted into a woman's womb. This method is often used by women who have difficulty conceiving a child naturally. In some cases, doctors make more embryos than are needed. Some people feel that these extra embryos should not be used for research purposes, either. They feel that if the embryos could at any time become living humans, it is immoral to use their stem cells.

Those who support stem-cell research point out that aborted fetuses and extra embryos from in vitro fertilization would be destroyed anyway. They believe that the potential medical breakthroughs that may arise from the use of embryonic stem cells could be of enormous benefit to millions of people.

ARSENIC AND HEART DISEASE

Recent research has shown that the ingestion of arsenic can lead to heart attacks, strokes, and diseased arteries in the extremities. *Arsenic* is a naturally occurring, extremely poisonous element that can be found in the water of some wells. Each year, millions of people around the world are exposed to arsenic in their well water. People in Asia are especially at risk.

Arsenic and other toxins may speed up atherosclerosis in arteries to the brain. In addition, arsenic may increase the risk of other blood vessel diseases.

AIR POLLUTION AND THE CIRCULATORY SYSTEM

Studies have shown that air pollution can have a negative effect on blood vessels. Ozone and other airborne pollutants can cause arteries and veins to *constrict,* or close, by 2 to 4 percent, reducing the flow of blood through the vessels. In healthy people, this tiny constriction may have no effect at all. However, for people already suffering from heart disease, the constriction might cause a heart attack or other serious problem.

Those who live in city areas are at a higher risk for some circulatory diseases. That's because they have been exposed for years to soot and dust from factories, cars, power plants, and other pollution sources. In the United States, people have long recognized that healthy air can lead to a healthier population. The 1990 Clean Air Act was one of the first federal attempts to limit air pollution.

The Environmental Protection Agency (EPA) recognized Denver, Colorado, as the first major metropolitan area in the United States to comply with the 1990 Clean Air Act.

HEALTH CONDITIONS AND LIFESTYLE CHOICES

Some circulatory problems may be caused or worsened by diet and lifestyle choices. For example, there is a well-founded link between heart disease and high-fat diets, obesity, and lack of exercise. Smoking is also a particularly dangerous choice for many people.

One potentially deadly disease that is worsened by bad diet and lifestyle choices is *cardiovascular disease* (CVD). CVD, also called heart disease, is any heart disorder caused by a reduced supply of blood to the heart. There are many types of CVDs, including coronary artery disease, heart attack, and stroke.

CVD is a serious problem in the United States. According to the American Heart Association (AHA), nearly 62 million Americans currently suffer from one or more types of heart disease. CVD is the number one killer of people in the United States.

Fast Fact

Nine out of ten adult smokers began smoking when they were teens.

Some CVD Statistics

Check out these stats from the AHA.

- On average, one American dies every thirty-three seconds as a result of CVD.
- CVD claims as many lives each year as the next seven leading causes of death combined.
- About one-third of all CVD-related deaths are people under the age of seventy-five.
- Of all people killed by CVD, 53.5 percent are women.
- Between 1989 and 1999, the rate of deaths due to CVD rose by 2.1 percent.

This man is suffering the effects of a heart attack.

What's in Cigarette Smoke?

Here are just a few of the hundreds of gases and toxins that cigarette smokers are inhaling:

- *Acetone* is a chemical that is used in nail polish and the production of explosives.
- *Ammonia* is a highly toxic chemical that is used in cleaning products.
- *Arsenic* is a metallic element that is used to make rat poison.
- *Benzene* is a chemical that is used to make dyes, detergents, plastics, and varnishes.
- *Butane* is a substance that is used to make lighter fluid and is added to petroleum gas.
- *Carbon monoxide* is the same gas that is expelled from car exhausts. In large quantities, it is poisonous.
- *Formaldehyde* is a chemical that, in liquid form, is used to preserve dead bodies.
- *Hydrogen cyanide* is the gas that is used to execute people in gas chambers.
- *Nicotine* is a highly addictive, poisonous substance that is used to make pesticides.
- *Tar,* the same sticky substance that is used to make roads, collects on the inside of smokers' lungs.

SMOKING

Smoking has been linked to many different types of heart disease. In some cases, smoking tobacco products can double the chances that a person may get a certain condition. Smoking reduces the amount of oxygen to the heart and other parts of the body. It can increase the heartbeat, raise blood pressure, and damage blood vessel linings. It can also raise cholesterol levels, increasing the risk of blood clots and other problems.

Secondhand smoke can also cause serious medical problems. Secondhand smoke is the smoke and chemicals that nonsmokers ingest from other people's tobacco smoke. According to the American Lung Association (ALA), secondhand smoke is responsible for as many as 62,000 heart-disease deaths annually.

Some people are unable to avoid breathing in the fumes of others. In a 1996 survey more than 37 percent of nonsmokers surveyed said that they were exposed to secondhand tobacco smoke at home or in the workplace. A more recent study published in 2003 shows fewer Americans being exposed to secondhand smoke. Any exposure, however, is still dangerous. The risk of coronary artery disease increases by up to 30 percent when nonsmokers inhale secondhand smoke.

Some Smoking Stats

Check out these smoking stats from the AHA.

- About one out of every five deaths from cardiovascular disease can be attributed to smoking.
- In 1990, 3 million people around the world died of smoking-related illnesses. The World Health Organization (WHO) expects this number to rise to 10 million by 2030.
- Just one year after quitting smoking, a person's risk of cardiovascular disease drops by 50 percent.
- In 1996, about 15 million kids under the age of eighteen were exposed to secondhand smoke in the home.

CORONARY ARTERY DISEASE

Coronary artery disease (CAD), sometimes called coronary heart disease, is caused by a narrowing of the arteries that bring blood to the heart. This narrowing is often the result of atherosclerosis. Less blood also means less oxygen into the heart. This can lead to a heart attack, along with permanent heart damage.

CAD is one of the most common types of heart disease in the United States today. An estimated 7 million Americans have CAD. Each year, 500,000 of them die from CAD-related heart attacks.

People at risk from CAD are those who smoke, eat a diet high in cholesterol and calories, or are not physically active. Other conditions, including diabetes, hypertension, and stress, can also heighten the risk. African American men aged fifty-five and older are also at increased risk of CAD.

The condition develops slowly, usually without causing any pain or discomfort in the early stages. People suffering from CAD may not even realize that they have the disease for many years. Most people learn that they have CAD only after they suffer a heart attack. For this reason, the condition is often called "the silent killer."

Those who do have symptoms may experience *angina* (an-JYE-nuh), which is severe, crushing pain behind the breastbone. Angina, although not a heart attack, is a warning sign that the heart is not getting the blood and oxygen it needs. Angina may be triggered by exercise, stress, cold weather, or any number of other factors. Another symptom is shortness of breath. The final symptom of CAD is a heart attack.

To treat CAD, doctors may use drugs to widen the arteries. Unfortunately, the drugs can also cause a slowing of the heart muscle by lowering blood pressure.

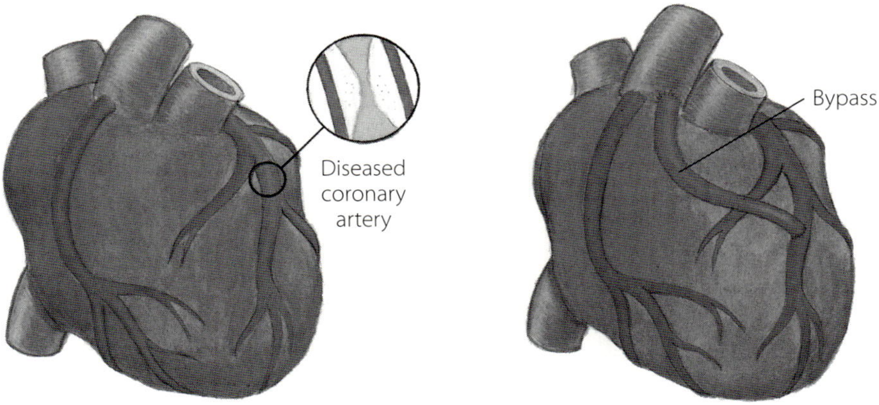

This diagram shows the effects of coronary artery disease on the left and a heart bypass on the right.

HIGH BLOOD PRESSURE

High blood pressure, also known as hypertension, is another serious health problem. About 50 million Americans currently suffer from high blood pressure. Normal blood pressure can be anywhere from 100/60 to 120/80. High blood pressure is anything above 140/90.

High blood pressure puts extra stress on the heart and blood vessels. It also strains the lungs, kidneys, and brain and may eventually damage these organs. High blood pressure can lead to heart attack, heart failure, stroke, arteriosclerosis, and other conditions.

Symptoms of high blood pressure include headaches, blurred or double vision, abdominal or chest pain, and shortness of breath. The exact cause of high blood pressure is unknown in 90 to 95 percent of all cases. Doctors do know, however, that many health conditions or lifestyle choices contribute to high blood pressure. Smoking, obesity, and high sodium diets are just a few factors that can raise blood pressure to an unhealthy level. Recent studies by the AHA have also linked high blood pressure with such socioeconomic factors as low income levels, low educational levels, and the quality of a person's neighborhood.

While most of the people suffering from hypertension are adults, children may sometimes be affected by the condition. Although the incidence of high blood pressure in children is low, some of them are at risk. Children who are obese, lack exercise, have a parent with high blood pressure, smoke or inhale secondhand smoke, or eat lots of salty foods are more likely to have high blood pressure.

Treating high blood pressure is often a question of making healthy lifestyle changes. A proper diet, regular exercise, quitting smoking, and avoiding alcohol can all help. Doctors may also prescribe medication to lower blood pressure.

High Blood Pressure Stats

Check out these stats about high blood pressure from the AHA.

- Between 1989 and 1999, the number of deaths as a result of high blood pressure increased by 46 percent.
- One out of every four adult Americans suffers from high blood pressure.
- Over 31 percent of all people with high blood pressure are unaware that they have it.
- Between the ages of twenty and fifty-four, men are more likely than women to have high blood pressure. After the age of fifty-five, this changes, and women have the higher risk.
- The southeastern region of the United States has the highest rate of high blood pressure.

Low Blood Pressure

Low blood pressure, also known as hypotension, is not considered a medical condition until it becomes low enough to cause health problems, such as dizziness and fainting. Blood pressure may become low as a result of infection, heart attack, certain medications or drugs, and other conditions or diseases.

HEART ATTACKS

Also known as a myocardial infarction (mye-oh-CARD-ee-ul in-FARK-shun), a *heart attack* occurs when the blood and oxygen supply to the heart muscle have ceased, causing part of the muscle to die. Blood clots that block an artery are most often the cause of heart attacks. Once part of the muscle has died, there is no way to repair the damage. The rest of the heart must work harder to compensate for the lost muscle.

Heart attacks are a common health condition in the United States. According to the AHA, more than 1 million people suffer a heart attack each year. Of these heart attack cases, 44 percent prove fatal. In fact, heart attacks are responsible for one out of every five deaths in the United States.

Smoking is just one of the factors that is known to increase the risk of heart attacks. Other factors include high blood pressure, high cholesterol levels, high fat diets, and obesity. By quitting smoking and eating healthy foods, people can reduce their chances of suffering a heart attack.

Symptoms of a heart attack include chest, back, and abdominal pain; pain that moves to the chest, arms, and shoulders or the neck, jaw, and head; shortness of breath; nausea and vomiting; fainting; sweating; anxiety; fatigue; and seizures. People with these symptoms should seek medical attention immediately. Some heart-attack victims, however, will experience no symptoms at all. This type of heart attack is sometimes known as a "silent attack."

Some heart attacks are more severe than others, depending upon how much of the muscle has died. To check for the amount of damage that a heart-attack patient has suffered, doctors test the patient's blood for certain enzymes. This is because the heart muscle releases these enzymes during a heart attack. One such enzyme is *creatine phosphokinase* (KREE-uh-teen fahss-foh-KYE-nase), or CPK, found in heart muscle. During a heart attack, CPK leaks into the bloodstream. Another enzyme that doctors check for in the bloodstream is *lactate dehydrogenase* (LAK-tayt dee-hye-DRAH-jeh-nase), or LDH. The more enzymes that doctors find in the blood, the greater the damage to the heart.

People who have suffered heart attacks are hospitalized for up to two weeks, depending upon the seriousness of their condition. During their stay in the hospital, doctors will constantly monitor their heart rate. Intravenous fluids and medicines help the patient heal and prevent further damage. In some cases, doctors may choose to perform surgery to open a blocked artery.

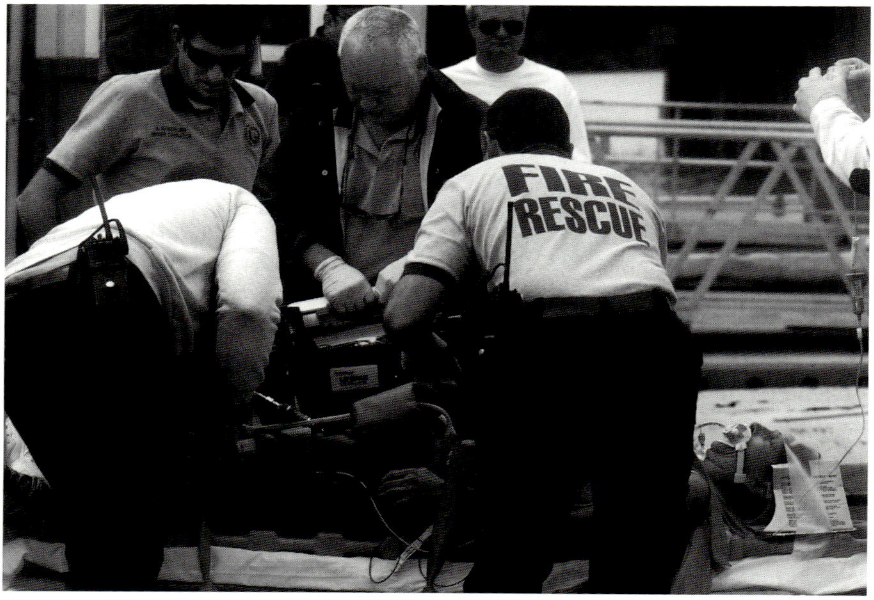

Paramedics work on a woman who has gone into cardiac arrest.

CARDIAC ARREST

Cardiac arrest occurs when the heart completely stops beating. Sudden cardiac death is usually the result. The most common cause of cardiac arrest is heart disease. Other causes may include heart defects, sudden bursts of adrenaline during physical activity, blows to the chest, and the use of certain legal and illegal drugs. Legal drugs that can cause cardiac arrest include certain heart medications. Illegal drugs that can cause cardiac arrest include cocaine and inhalants. Electrocution, drowning, and other traumas can also cause cardiac arrest. In some cases, however, the cause remains unknown.

Cardiac arrest occurs suddenly and usually without warning. A person suffering from cardiac arrest will stop breathing, become unresponsive, and lose pulse. Because patients often die just minutes after these symptoms begin, it is important that medical help be sought immediately. A cardiac-arrest victim is more likely to survive if he or she is immediately given cardiopulmonary resuscitation (CPR) and an electric shock to restart the heart.

In recent years, the number of young women between the ages of fifteen and thirty-four experiencing sudden cardiac death has increased dramatically. Although they are not sure why, health professionals point to smoking, obesity, and lack of physical activity as possible factors.

HEART FAILURE

Heart failure occurs when the heart no longer pumps blood properly. About 5 million Americans suffer from this condition. That number is expected to increase dramatically over the next few years as the number of Americans over the age of forty also increases. Heart failure is a serious medical condition. About two-thirds of all patients diagnosed with the condition die within five years.

Heart failure usually does not occur suddenly. Instead, the heart becomes weaker and weaker over the years, until it stops working altogether. Although most hearts do lose pumping efficiency with age, heart failure occurs when this condition is much more serious and pronounced than normal. This may be the result of an illness or another condition that has damaged the heart, including rheumatic fever or a heart attack.

There are two main types of heart failure: systolic and diastolic. *Systolic* heart failure is a failure of the left side of the heart. The majority of heart failures occur on the left side of the heart. In such situations the heart loses its ability to contract and pump blood through the body. In some cases, fluid may leak back into the lungs, causing a condition called *pulmonary congestion. Diastolic* heart failure is a failure of the right side of the heart. The heart loses its ability to relax and fill with blood. Fluid sometimes builds up in the feet, ankles, and legs.

Two of the most common causes of heart failure are high blood pressure and CAD. Other causes include anemia, congenital heart defects, and *hyperthyroidism* (hye-per-THYE-roid-iz-um), which is an overactive thyroid (THYE-royd) gland. People increase their chances of suffering from heart failure by not eating properly, smoking, drinking excessive amounts of alcohol, or using anabolic steroids for long periods of time. Obesity and diabetes also increase the risk.

Symptoms of heart failure include breathlessness, fatigue, coughing, and a buildup of fluid in the lungs and other tissues. This fluid buildup is called *edema* (eh-DEE-muh). Although there is no cure for heart failure, doctors can treat the disease with certain drugs that help the heart work better. Recommended lifestyle changes include losing weight, quitting smoking, and not drinking alcoholic beverages.

Sometimes surgery is necessary. Operations to repair heart failure include heart-valve replacements, artery bypasses, and heart transplants.

DEEP VEIN THROMBOSIS

Another condition that smokers are at a higher risk for is *deep vein thrombosis,* (DVT). A *thrombosis* (thrahm-BOH-siss) is the formation of a clot, or *thrombus,* in a blood vessel. When the thrombosis occurs in a deep vein, it can be painful and potentially life threatening. The blood clot blocks the flow of blood back to the heart. It can cause tissue damage and gangrene. A clot that breaks free from the vein in the leg and travels to another part of the body is called an embolism. If the embolism travels to the lungs, death can result.

DVTs are common conditions. Each year, about 5 million Americans suffer from this type of venous disorder. The condition usually occurs in people over the age of forty. DVTs can be caused by prolonged inactivity, injury, or surgery. In addition, obesity, high cholesterol levels, and hypertension can all increase the risk of DVT.

Symptoms of DVT include pain, swelling, enlargement of the superficial veins in the area, and warm and discolored skin. Doctors may treat people with bed rest, elevation of the affected area, and blood-thinning drugs, called *anticoagulants* (an-tee-koh-AGG-yoo-lentz), to break up the clot. In some cases, clot-dissolving medicine is injected into the body near the thrombus site.

STROKES

Also known as a “brain attack,” a stroke occurs when the blood supply to the brain is cut off. Brain cells and tissue begin to die. Strokes can cause serious damage to the brain, leading to paralysis, as well as problems thinking and speaking. This condition is often caused by carotid artery disease, in which the carotid artery becomes thick and hard.

New Ways to Fight Strokes

A backpack-sized invention may save people around the world from suffering strokes. The device, called SonoSite 180, is a small, portable ultrasound machine that can be used to check for carotid-artery blockage. This type of blockage is a major cause of stroke. The SonoSite can be easily transported to small communities and clinics, unlike the big, bulky ultrasound machines in hospitals and laboratories. Currently, each machine costs as much as $20,000 for a hospital to buy.

There are two major types of strokes: ischemic and hemorrhagic. An *ischemic* (iss-KEEM-ik) *stroke* is caused when a blood clot blocks the supply of blood in a blood vessel in the brain. Eight out of ten strokes are ischemic. A *hemorrhagic* (hem-or-AJ-ik) *stroke* is caused when an aneurysm bursts and blood leaks into the brain.

Stroke is a serious problem. This deadly condition is the third leading cause of all deaths in the United States. It is also one of the chief causes of long-term disability. According to the AHA, an American suffers a stroke every fifty-three seconds; one American dies as a result of a stroke every 3.1 minutes.

As with many other circulatory problems, people who smoke are at increased risk of this condition. In fact, smoking doubles the chances of stroke. People with high blood pressure, heart disease, diabetes, and high cholesterol are also more likely to suffer a stroke. Those who drink alcohol heavily are at especially high risk for hemorrhagic stroke.

Symptoms of a stroke occur suddenly. They may include numbness or weakness on one side of the body; confusion; difficulty seeing, walking, and talking; dizziness; and severe headache. It is important for people with stroke symptoms to seek medical help immediately. Successful treatment often depends upon how quickly a patient gets help. The longer a person's brain is deprived of blood, the more likely that person is to have serious, long-term complications as a result.

Doctors treat ischemic strokes with blood thinners and platelet inhibitors designed to dissolve clots that may be blocking the flow of blood to the brain. To work effectively, the drug must be given to patients within three hours of the stroke's onset. Hemorrhagic strokes are life-threatening conditions. To treat this type of stroke, doctors work to stop the bleeding and minimize any damage from the condition.

However, health professionals say that the best way to treat a stroke is to take steps to prevent it from ever happening. This means eating a low-fat, low-cholesterol diet; not smoking; taking care of high blood pressure; and trying to stay as healthy as possible.

Smokeless Tobacco

Smokeless tobacco, sometimes called chew or spit, has become increasingly popular among teen boys. Many believe that by chewing tobacco instead of smoking it, they are avoiding the health problems associated with cigarette smoking. This is not true.

Chewing tobacco carries its own serious health risks. It can increase the risk of such circulatory conditions as heart attacks, angina, atherosclerosis, high blood pressure, and blood clots. In addition, chewing tobacco increases a person's risk for oral cancer by fifty times.

ANEMIA AND DIET

People's diets may increase their risk for certain circulatory disorders. Anemia, for example, may be caused by nutritional deficiencies in the diet, especially a lack of iron. Anemia is the most common type of blood disorder, caused by a lack of red blood cells in the body. As a result, people with anemia don't get enough oxygen in their bodies. Symptoms of the disease include pale skin, fatigue, dizziness, and rapid heartbeat.

There are a number of different kinds of anemia. Some are caused by dietary deficiencies, while others are caused by blood loss, infection, and other health conditions. The most common type of anemia in children is *iron deficiency anemia,* which results when children get too little iron in their diets.

Pernicious (pur-NISH-uss) *anemia* is caused by a lack of the substance needed to process vitamin B12. Like iron, B12 is a nutrient that is essential to the normal production of blood.

One type of anemia can be caused by environmental factors. *Aplastic* (ay-PLASS-tik) *anemia* is caused by exposure to chemicals, other toxins, and some types of medications. With aplastic anemia, the body's bone marrow can no longer produce enough blood cells. This type of anemia is rare.

Because anemia can be the result of so many different factors, doctors must determine the cause of anemia before being able to treat it successfully. When anemia is caused by a lack of iron, doctors may recommend that patients take an iron supplement or eat foods that are rich in the mineral. Patients may also need a blood transfusion or bone marrow transplants to correct the problem. In severe cases, the spleen may be removed from the body. The *spleen* is an organ responsible for filtering worn-out red blood cells from the body. When the spleen is not working properly, however, it may produce abnormal antibodies that can lead to anemia. The spleen is not essential for life, but people without this organ are at a higher risk for infection than other people are.

TREATING HEART DISEASE AND HEART ATTACKS

There are a number of ways that doctors treat people who have heart disease or who have suffered heart attacks. Numerous medicines can help people with heart problems, including the following:

- *Diuretics* (dye-yoor-ET-iks) are drugs that increase the removal of water from the body. As a result, blood pressure decreases.
- *Beta blockers* slow the heart rate and reduce the strength of the heart muscle's contractions. As a result, blood pressure decreases.
- *Channel blockers* stop calcium from entering the blood vessel walls. This helps the blood vessels remain dilated and healthy.
- *Angiotensin-converting* (an-jee-oh-TEN-sin) *enzyme* (ACE) *inhibitors* block the production of an enzyme that makes blood vessels constrict. This helps blood vessels remain dilated and lowers blood pressure.

Cardiopulmonary resuscitation (kar-dee-oh-PUL-min-er-ee ree-suss-ih-TAY-shun), or CPR, is mouth-to-mouth breathing performed on those suffering from serious heart attacks or cardiac failure. The person performing CPR helps the patient breathe by forcing oxygen into his or her lungs. At the same time, the person puts pressure on the chest to help blood continue to circulate to the heart. CPR training is an important emergency medical technique that can keep the lungs and heart going until help arrives. Anyone can learn to perform CPR.

Those suffering from some types of heart disease may be treated with *defibrillation* (dee-fib-ruh-LAY-shun). A machine called a *defibrillator* (dee-FIB-ruh-lay-tor) is used to apply an electric shock to the chest. The electric shock helps normalize the beating of the heart. People with certain types of heart disease may have a small defibrillator implanted into their chests to control their heartbeat. These implanted defibrillators are known as *pacemakers.*

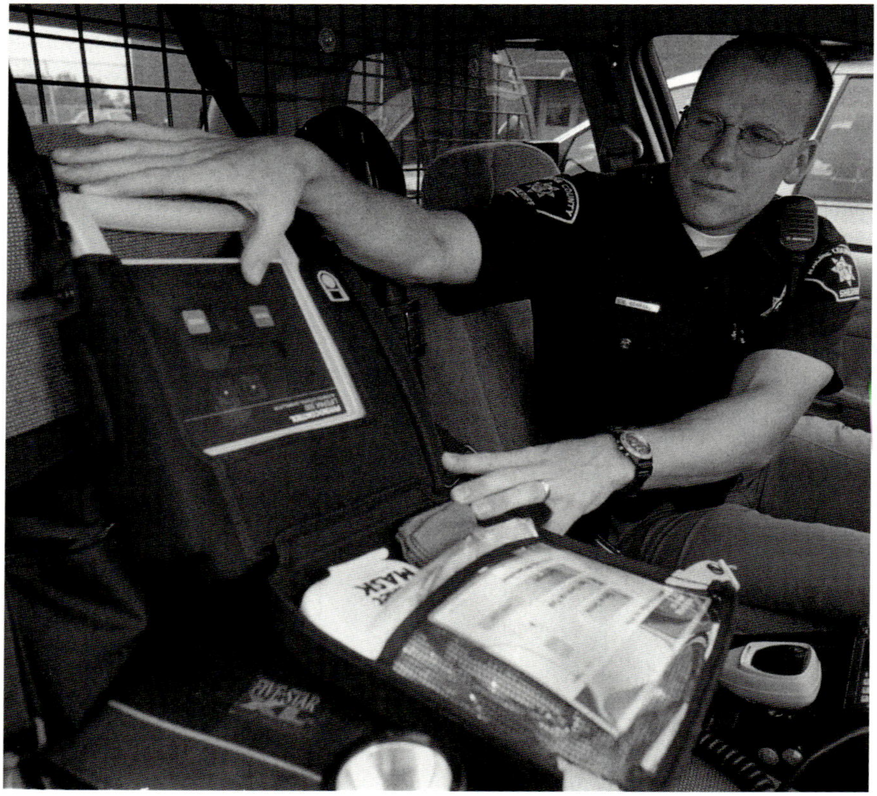

Portable defibrillators are used by emergency personnel such as this police officer from Racine, Wisconsin.

There are several medical procedures that doctors use to treat heart disease. An *angioplasty* (AN-jee-oh-plass-tee) is a procedure in which a catheter is inserted into a blood vessel, usually in the groin or arm. The catheter is then threaded to a blocked artery. Once the catheter is in place, a balloon at its tip is inflated. This balloon cracks or flattens any plaque that is blocking the artery. In some cases, a *stent,* or small wire tube, is inserted to keep the vein open. An angioplasty is one of the least invasive ways to treat heart disease surgically. Unfortunately, about one-third of all blockages reoccur after an angioplasty is performed. When this happens, a bypass may be necessary.

During a *coronary artery bypass,* surgeons make a pathway around the blocked blood vessel. A piece of vein or artery from elsewhere in the body, usually the leg or the chest, is used to create the alternate route. During the surgery, the heart must be stopped so that surgeons can work on it. The patient is kept alive by a special machine called a *heart-lung machine.* During the surgery, doctors may bypass one or more arteries.

A *heart transplant* is an operation in which the patient's own heart is removed and replaced with a healthy new heart. The new heart is taken from a donor who has been declared brain dead but has been kept alive with life-support machines. This type of treatment is usually reserved for patients who have not responded well to other treatments. Heart transplants are the fourth most common type of organ transplant in the United States. Each year, more than 2,000 heart transplants are performed.

INNOVATIONS IN HEART TREATMENT

In recent years, doctors have been using new techniques to treat people suffering from heart disease or heart attacks. *Gene therapy* is one example. Scientists are studying the use of genes to stimulate the growth of new blood vessels in the body. Doctors may even be able to inject genetic material directly into the heart to restore blood flow.

In *laser revascularization* (ree-vass-kyoo-lur-ih-ZAY-shun), a laser is inserted into the heart through a catheter. Laser bursts create tiny channels in the heart muscle. Scientists think that the laser bursts stimulate the growth of new blood vessels.

In a *cardiomyoplasty* (kar-dee-oh-MYE-oh-plass-tee), muscles are taken from a person's back or abdomen and wrapped around the heart. The new muscles are expected to increase the heart's pumping power.

VIRUSES AND BACTERIA

Many circulatory problems are caused by viruses, bacteria, and other harmful organisms that invade the body. One viral condition is *myocarditis* (mye-oh-kar-DYE-tiss), an inflammation of the heart muscle. Myocarditis can cause permanent damage to the heart muscle and may even result in death. Another circulatory problem often caused by viral infection is *pericarditis* (perr-ih-kar-DYE-tiss). This condition is an inflammation of the pericardium, the protective membrane that surrounds the heart. Pericarditis may result in a buildup of fluid around the heart. Severe cases may be life threatening and may require surgery.

One circulatory condition caused by bacteria is *endocarditis* (en-doh-kar-DYE-tiss). Endocarditis is an infection of the heart muscle, chamber, or valves. The infection often results when germs enter the body during surgery. People with congenital heart valve defects are at high risk for endocarditis.

INFECTIOUS MONONUCLEOSIS

Infectious mononucleosis (mahn-oh-noo-klee-OH-siss), or mono, is caused by the Epstein-Barr virus (EBV). This virus, a member of the herpes family, is one of the most common human viruses. Most people in the world will be infected with the virus at some point in their lives. It is estimated that 95 percent of all Americans between the ages of thirty-five and forty have already been infected with EBV.

When people are first infected with EBV, they may not even notice it. The immune system usually does an excellent job of destroying the virus. Problems may arise, however, when adolescents and young adults are infected with EBV. Between 35 and 50 percent of all young people who get the virus then may develop mono. Symptoms of mono include fever, sore throat, swollen lymph nodes, and fatigue. The disease is usually not serious, and symptoms normally disappear in two months or less without treatment.

Mono is a contagious disease. It is transmitted from one person to another through saliva. As a result, mono has earned the nickname "the kissing disease." Coughing, sneezing, or sharing a glass can also spread the disease through infected saliva. However, people who have already been infected with EBV don't have to worry: They are immune to mono.

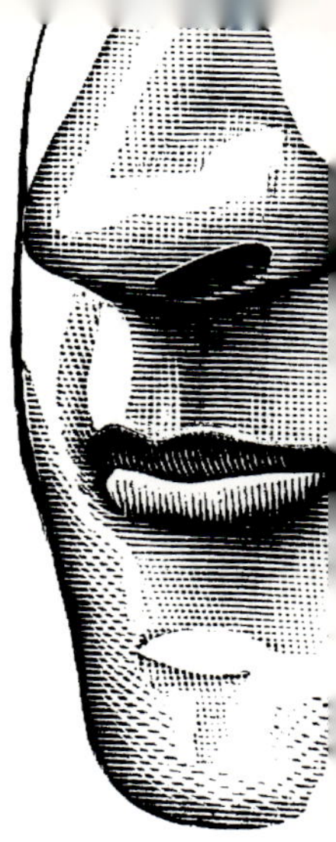

4

THE RESPIRATORY SYSTEM

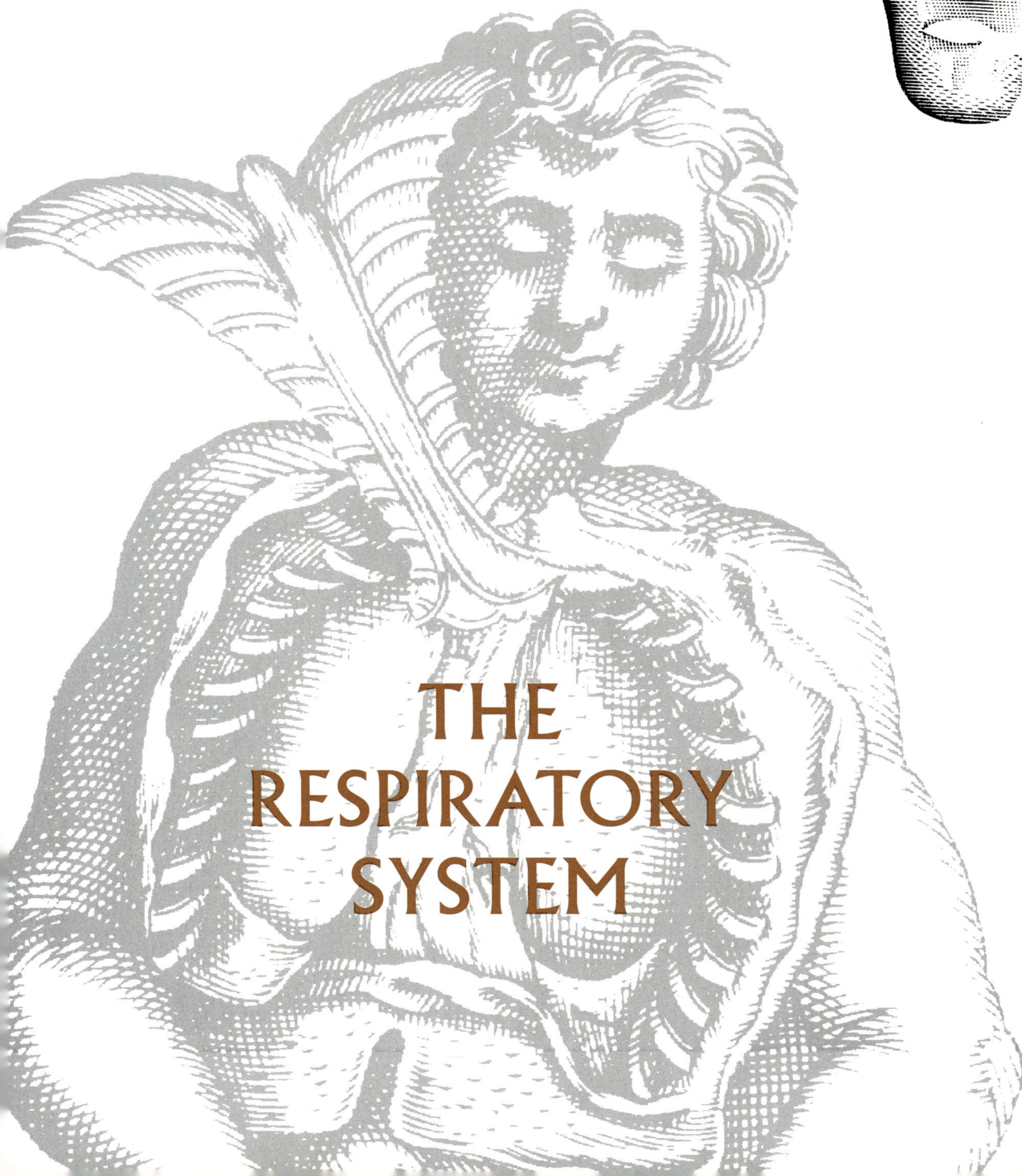

The *respiratory* (RESS-pur-uh-tor-ee) *system* controls the body's breathing which is how people get the oxygen that is essential to all life. The process begins when people breathe in oxygen and other materials from the air around them. When they breathe out, they exhale waste gases, especially carbon dioxide, from the body.

The process of breathing in oxygen and breathing out carbon dioxide is called *respiration.* Respiration is unconscious, or automatic. Even when people are asleep, the brain and nervous system control breathing.

Another type of respiration is *cellular respiration.* Cellular respiration, also known as metabolism, takes place in every cell in the body. It is the process through which cells convert carbohydrates into energy that the entire body can use. During the process, the cells use oxygen and release carbon dioxide.

U.S. free-diving champion Audrey Mestre Ferreras takes a last deep breath before plunging to a depth of 426.5 feet (130 meters) to set a new record. Ferreras had to hold her breath for more than two minutes to achieve the record.

THE UPPER RESPIRATORY TRACT

The respiratory cycle begins when a person takes a breath of air into the nose or mouth. Air that passes through the mouth and nose is warmed and moistened. Air that passes through the nose is cleaned, as well. Special cells inside the nasal cavity, called *goblet cells,* make mucus, which traps dust particles. *Cilia* (SIL-ee-uh), special hairs in the nasal cavity, move dust and mucus toward the nostrils to be sneezed or blown out.

From the nose and mouth, air continues down the throat, or *pharynx* (FAAR-inks). The pharynx is lined with mucous membranes that continue the cleansing process started in the nasal cavity.

Together, the nose, nasal passages, and upper pharynx form what is known as the *upper respiratory tract.* Many types of respiratory diseases and infections affect the upper respiratory tract. One of the most common upper respiratory conditions is a cold.

THE LOWER RESPIRATORY TRACT

Air next enters the *lower respiratory tract.* From the pharynx, the air enters a tube called the *larynx* (LAAR-inks). The top of the larynx is covered by a special flap of skin called the *epiglottis* (eh-pih-GLAHT-tiss). When a person eats, the epiglottis covers the larynx and prevents food from entering. The voice box is also found in the larynx. Air vibrations on the voice box allow humans to speak and make other noises.

From the larynx, air enters the *trachea* (TRAY-kee-uh), or windpipe. The trachea is the main passageway to the lungs. Eventually, the trachea splits into two different branches, or *bronchi* (BRON-kye). One bronchus leads to the left lung, while the other leads to the right lung.

The lungs are the key organs of the respiratory process. It is there that blood is supplied with oxygen for the rest of the body's cells and

Fast Fact

While the right lung has three lobes, or segments, the left has only two. This allows room for the heart.

tissues. The lungs also filter out carbon dioxide and other waste matter from the blood. How does this happen? In each lung, the bronchi branch out into smaller tubes, called *bronchioles* (BRON-kee-ohlz). The bronchioles all end in tiny air sacs called *alveoli* (al-VEE-uh-lye). Here, the thin walls of the alveoli allow oxygen to pass into the bloodstream while carbon dioxide is filtered out. Each lung has 300 million to 400 million alveoli.

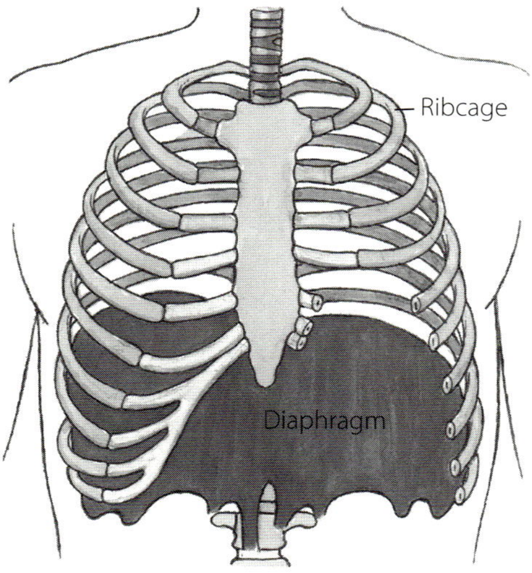

The lungs take up nearly all of the chest cavity. They are protected from damage by the ribs. Each lung is covered by a protective membrane called the *pleura* (PLOOR-uh). The pleurae keep the lungs from being damaged by the ribs during breathing.

The muscle that powers respiration is the *diaphragm* (DYE-uh-fram), a dome-shaped barrier that separates the chest from the abdomen. When a person inhales, the diaphragm *contracts,* or shrinks. This allows air to enter the lungs. When a person exhales, the diaphragm relaxes and expands. This forces air out of the lungs.

Fascinating Respiratory Facts

- An adult normally breathes about fifteen times every minute.
- When filled to capacity, the lungs can hold about 10 pints (4.7 liters) of air.
- Normal breathing is an automatic, or involuntary, process. The nervous system automatically regulates breathing depending upon the amount of oxygen and other gases in the air around you.
- The human body can survive for only about three minutes without air.
- The lungs of an infant are light pink in color. As humans age, however, the lungs darken and scar.
- The lungs of people who live in urban areas are generally darker than those who live in rural areas.
- A hiccup is an involuntary spasm of the diaphragm.

The Respiratory System Diagrammed

Here is an up-close look at some of the important parts of the respiratory system.

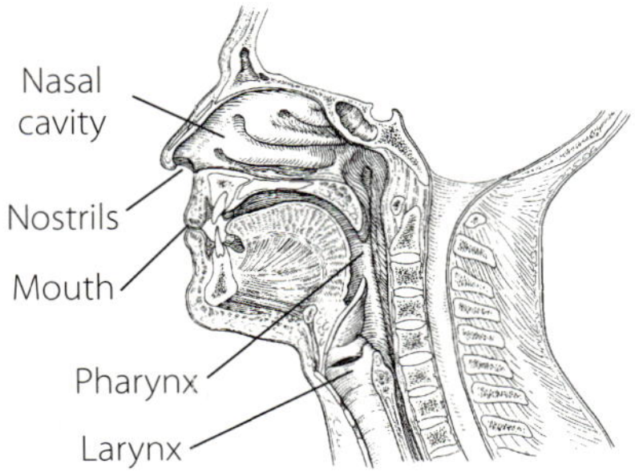

NOSE AND MOUTH—The respiratory cycle starts here when air is inhaled.

NOSTRILS—These two holes in the nose allow air in and out.

NASAL CAVITY—Inside the nasal cavity, air is moistened, warmed, and cleaned.

PHARYNX—The pharynx is a path for both food and air. The tonsils are located in the pharynx.

LARYNX—This tube connects the pharynx to the trachea.

TRACHEA—Also known as the windpipe, the trachea is located at the bottom of the larynx.

BRONCHI—These two branches funnel air into the lungs.

LUNGS—These cone-shaped, spongy organs are the chief respiratory organs.

BRONCHIOLES—As the bronchi become narrower, they branch off into these smaller tubes.

ALVEOLI—In these tiny air sacs at the end of the bronchioles, oxygen from the lungs moves into the blood. Waste gases leave the blood here, to be removed from the body during exhalations.

DIAPHRAGM—This is the main muscle involved in respiration.

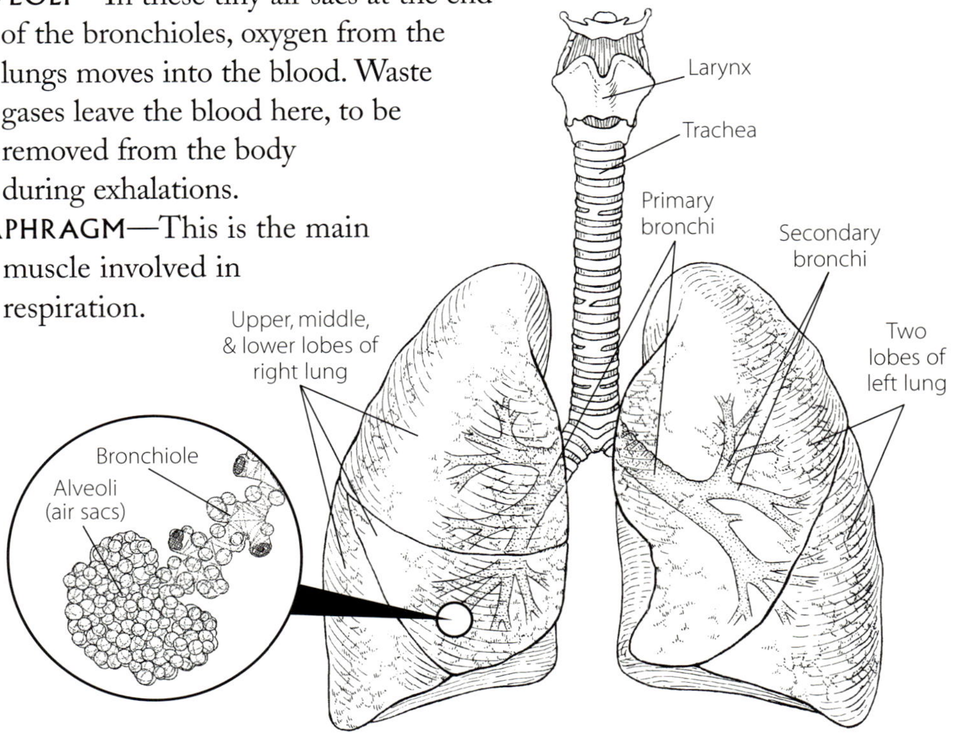

The Heimlich Maneuver

Sometimes, a piece of food or another object becomes caught in a person's trachea. If the object cannot be dislodged, it blocks the flow of air into the lungs. Without air, a person could suffocate and die. In 1973, Dr. Henry Heimlich (1920–) developed a procedure that uses the air in the lungs to expel the object. In the Heimlich maneuver, a person forces air out of the lungs by applying sudden pressure on the abdomen. In many cases, the rush of air from the lungs pops the object out of the trachea.

Dr. Henry Heimlich watches as lifeguards demonstrate how the Heimlich maneuver can save drowning victims at a Cincinnati swimming pool.

5

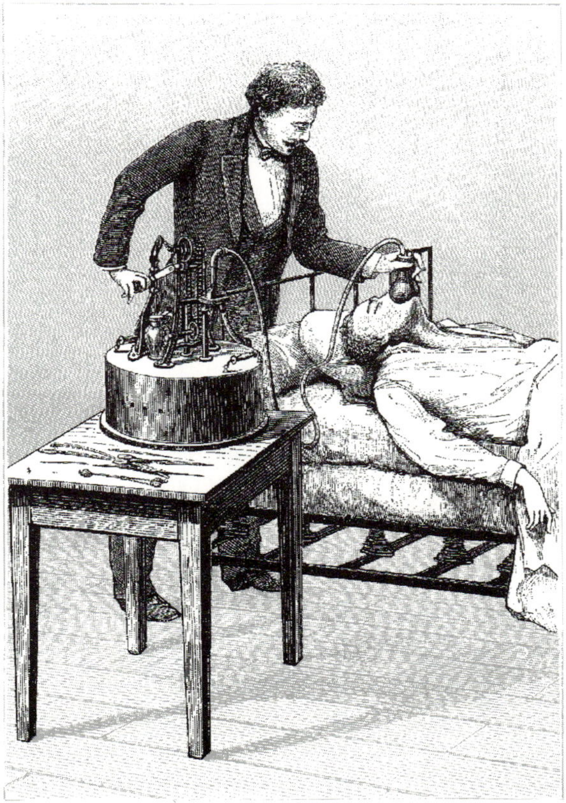

RESPIRATORY DISORDERS

Diseases and other medical conditions can affect any part of the body. The respiratory system is no exception. Respiratory disease inhibits a person's ability to breathe properly. Respiratory disease may also cause problems with oxygen and carbon dioxide levels throughout the body. As a result, the rest of the body is also affected, and other medical disorders may result.

CYSTIC FIBROSIS

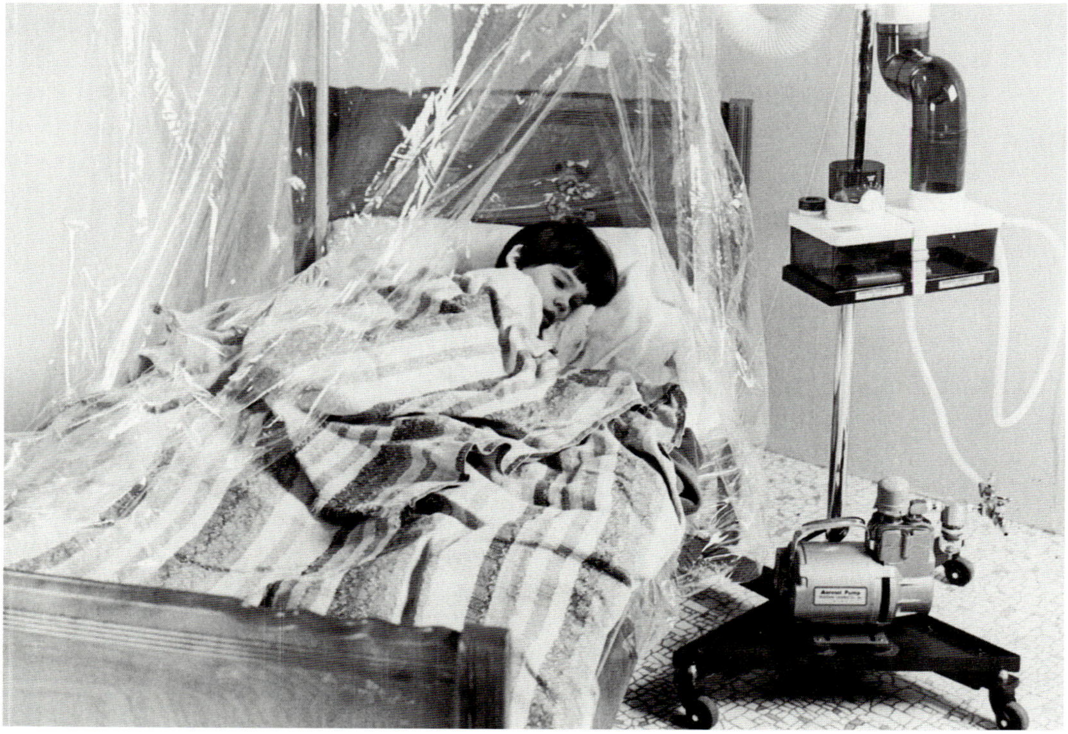

This boy suffers from cystic fibrosis. He sleeps beneath a plastic tent while a nebulizer delivers moist, medicated air for him to breathe to prevent his lungs from becoming clogged. Also, several times each day, he uses a face mask that helps him breathe.

Of the many diseases and disorders that can affect the respiratory system, some are congenital, or present at birth. One such disease is *cystic fibrosis* (SISS-tik fye-BROH-siss), or CF. CF is a fatal, hereditary condition in which mucous membranes in the respiratory tract produce abnormally thick mucus. This mucus collects in the lungs and intestines.

Symptoms of CF include frequent respiratory infections, especially pneumonia; difficulty breathing; persistent diarrhea; wheezing; and mucus-filled coughing. The disease often results in permanent lung damage. CF may also lead to other medical conditions, including malnutrition; sinusitis; nasal polyps, which are fleshy growths; pneumothorax (noo-moh-THOR-aks), or collapsed lung; diabetes; and liver disease.

CF is the most common inherited, life-threatening disease among young people in the United States. According to the National Heart, Lung, and Blood Institute, 2,500 babies are born with CF in the United States annually. More than 30,000 Americans are currently living with this serious condition.

CF is a *progressive* condition. This means that it gets worse over time. In the past, CF was always fatal in childhood. Today, thanks to better medicines and other treatments, children born with CF are living longer. The life expectancy for a person with CF is now about thirty years. People who die as a result of the condition usually die of lung disease.

DIAGNOSING AND TREATING CF

The most common test to see if a patient has CF is the sweat test. People with CF produce excessive amounts of salt when they sweat. Doctors use a special chemical and electric shock to make a section of the skin sweat. They then collect the sweat with gauze for testing. If the sweat sample has high levels of sodium and chloride, the patient probably has CF. Other tests used to diagnose CF are chest X-rays, tests of lung function, and sputum (phlegm) cultures.

There is no way to prevent or cure CF. To prolong a patient's life, doctors recommend treatment of the various conditions caused by CF. For example, physical therapy, exercise, and medications can break up mucus in the lungs. In some cases, doctors may perform a lung transplant.

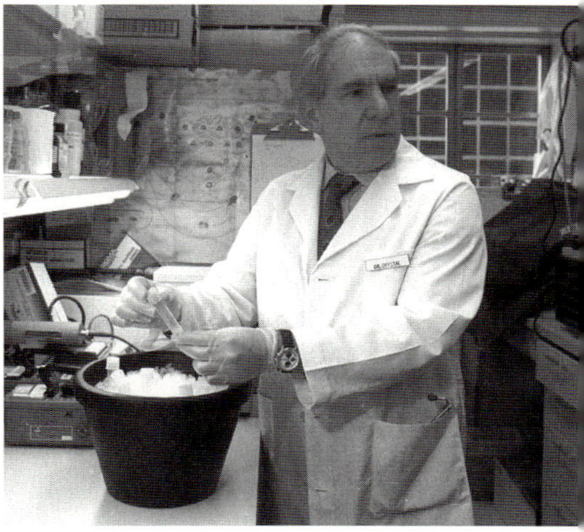

Dr. Ronald Crystal of the National Heart, Lung and Blood Institute in Bethesda, Maryland, is working on a nasal spray made from genetically engineered protein that breaks up the lung-clogging mucus of people suffering from cystic fibrosis.

Researchers are currently looking for ways to detect and control CF. Genetic testing can identify many people who carry the abnormality that causes this disease. After learning they have this trait, people can decide whether or not they want to have children.

Scientists are also looking for ways to treat or cure infants who are born with CF. One experimental treatment is gene therapy. Scientists hope someday to be able to introduce a gene into the body that would correct the abnormal gene that causes CF.

AUTOIMMUNE DISORDERS

Some respiratory diseases may be autoimmune disorders. Autoimmune disorders can affect any part of the respiratory system.

One respiratory condition that may be caused by a faulty immune system occurs when small lumps grow on the lungs. These lumps are made up of blood vessels, cells, and connective tissue. The condition is called *sarcoidosis* (sar-koy-DOH-siss). Another, more serious, autoimmune condition, *idiopathic* (ih-dee-oh-PATH-ik) *pulmonary fibrosis* (IPF), causes scarring and thickening of the lung walls. Eventually, this scarring may become so serious that the lungs are no longer able to get enough oxygen. Other autoimmune disorders that affect the lungs include lupus, scleroderma, and rheumatoid arthritis. These diseases can affect other body systems, as well.

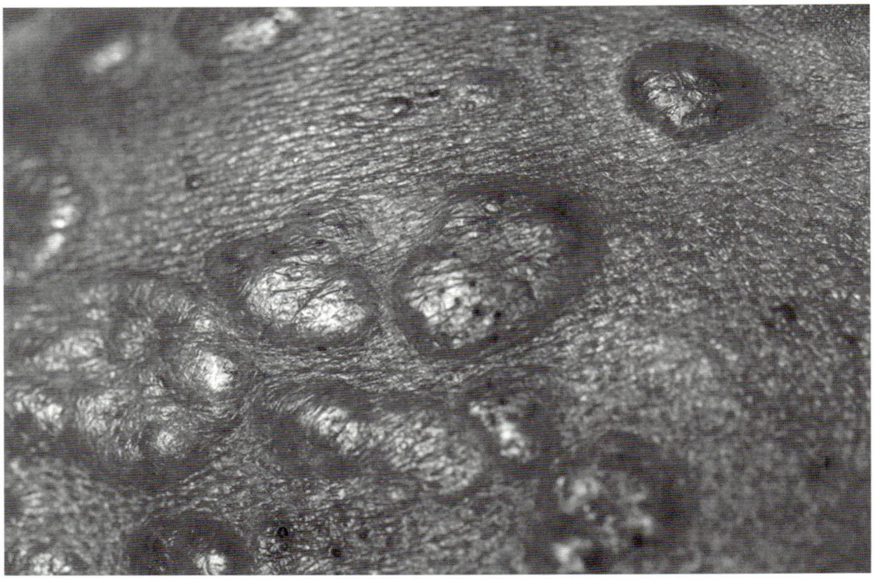

The lumps on the lungs caused by sarcoidosis are apparent in this photograph of lung tissue.

MISCELLANEOUS CONDITIONS

Respiratory diseases can be caused by any number of factors. Some are caused by other medical conditions. Some are caused by injury or trauma. In some cases, doctors may never know exactly what causes a particular respiratory condition.

One type of respiratory disorder that may be triggered by one of many different causes is *adult,* or *acute, respiratory distress syndrome* (ARDS). ARDS is a sudden failing of the lungs. This condition occurs when the lungs become inflamed and fluid builds up in the alveoli. Symptoms of ARDS include labored or rapid breathing, feelings of panic or anxiety, and low blood pressure. ARDS, sometimes known as stiff lung or shock lung, may be caused by traumatic injuries such as drowning, burns, or gunshot or knife wounds; infections; medical conditions, including shock or pneumonia; and drug overdose.

ARDS is a serious, life-threatening condition. Forty percent of people who suffer from this condition die. Those who recover may suffer permanent lung damage, memory loss, and difficulty thinking.

There is currently no way to prevent ARDS, but the condition can be treated effectively. Most people must be hospitalized and cared for in an intensive care unit (ICU). They may be intubated and hooked to a breathing machine. When a patient is *intubated,* a tube is passed through the nose or mouth into the trachea. The machine supplies oxygen to the body and helps the patient breathe.

DIAGNOSING RESPIRATORY PROBLEMS

Doctors use a number of means to diagnose respiratory problems. One way is to do a chest X-ray. X-rays can show an accumulation of fluid around the lungs. They can also help doctors diagnose lung cancers, collapsed lungs, tuberculosis, and other respiratory conditions. Another test is an *ultrasound.* Also called a sonogram, an ultrasound uses high-frequency sound waves to create a picture of the lungs and chest. Unlike an X-ray, radiation is not involved.

In a *bronchoscopy* (bron-KAH-scuh-pee), a tube is inserted through the nose, down the larynx, through the trachea, and into the bronchi of the lungs. Attachments can be put into the tubes to extract tissue samples.

To diagnose *pleurisy* (an inflammation of the pleura), doctors sometimes perform a *thoracentesis* (thor-uh-sen-TEE-siss). During this procedure, a needle is inserted through the chest and into the pleura to extract fluid for testing.

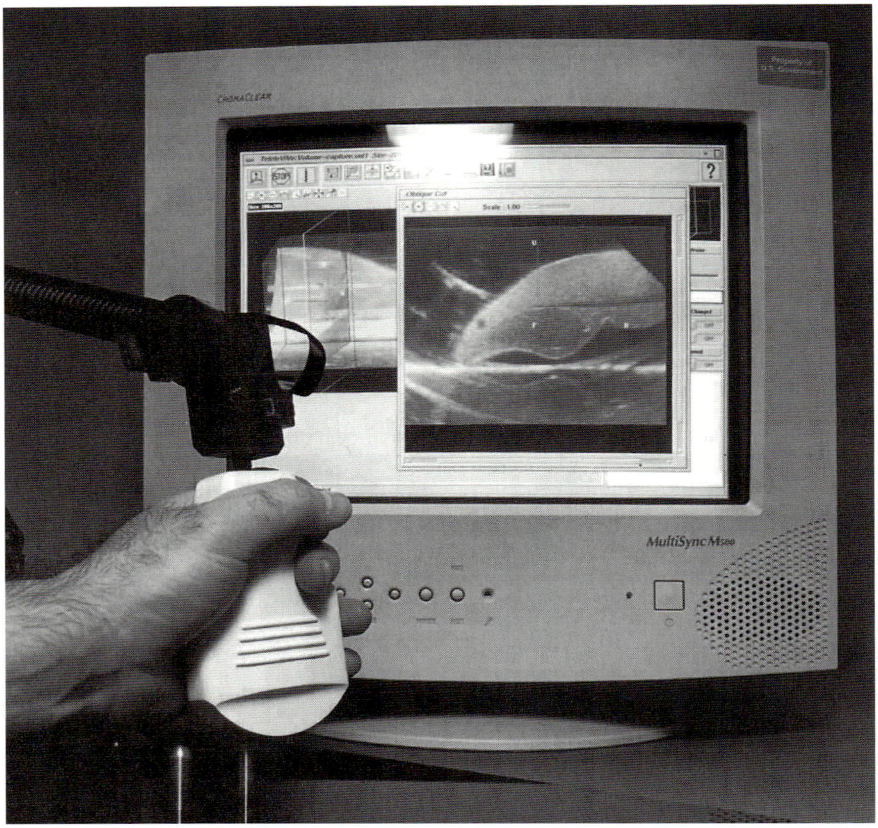

This remote ultrasound device was developed to diagnose problems in patients thousands of miles away. It transmits the ultrasound information to a remote location where a doctor can analyze it.

6

HOW THE ENVIRONMENT AFFECTS THE RESPIRATORY SYSTEM

Each day, we breathe more than 25,000 times. We inhale more than 2,600 gallons (9,880 liters) of air. Although most of what we inhale is oxygen and nitrogen, there are other particles floating in the air. These substances—bacteria, soot, smoke, chemicals, dirt—are all breathed into your body. Some of them may just irritate the lungs and respiratory system. However, other substances can cause serious health problems and may even be fatal.

AIR POLLUTANTS AND RESPIRATORY CONDITIONS

Substances that we inhale can seriously affect our health. Air pollution, for example, causes an increase in respiratory problems. It has also been linked to birth defects, heart attacks and other circulatory problems, and cancer. According to the Centers for Disease Control (CDC), as many as 120,000 people die each year as a result of exposure to air pollution.

Fast Fact

According to the ALA, 142 million Americans live in areas where the air that they breathe is unhealthy. The ALA ranked Los Angeles, California, as the city with the worst air quality.

People in city areas are especially at risk for pollution-related respiratory problems. Each day, they breathe in more soot, dust, auto emissions, and other pollutants than do those who live in rural areas. Research has shown that although some people in polluted areas may not have any symptoms of respiratory conditions, they may still be suffering permanent lung damage.

One of the most dangerous of all air pollutants is *ozone.* Ozone is a poisonous form of oxygen. It is created when an electrical charge passes through oxygen. Ground-level ozone is released into the air through motor-vehicle exhausts and industrial emissions. It can irritate and inflame the respiratory system. Ground-level ozone increases the risk of asthma, respiratory infections, and other breathing problems.

Another type of air pollutant that is a serious health problem is tobacco smoke. Most people who choose to smoke understand that doing so will damage their health. What they may not understand, however, is that their choice also affects those around them. People exposed to secondhand smoke run a higher risk of respiratory trouble than do nonsmokers who are not exposed to secondhand smoke.

LUNG DISEASE

Lung disease is any condition or disorder that affects the lungs. In the United States, lung disease is quite common, affecting more than 30 million Americans.

There are many different types of lung diseases. Doctors often classify these diseases by the part of the respiratory system they affect.

- *Diseases of the airways* affect the flow of air through the air passages. Such conditions include asthma, bronchitis, and emphysema.
- *Diseases of the interstitium* (in-ter-STIH-shum) affect the space between the tissues of the lungs. Such conditions include sarcoidosis and pulmonary fibrosis.
- *Disorders of gas exchange and blood circulation* affect the way gases and fluids circulate through the respiratory system. Such conditions include respiratory failure and ARDS.
- *Disorders caused by unusual atmospheric pressure* affect the lungs when a person travels somewhere where the air pressure is different than what he or she is used to. Such conditions include acute mountain sickness and high-altitude pulmonary edema.
- *Disorders of the pleura* affect the lung's protective membrane. Such conditions include pleurisy and pneumothorax.

ASTHMA

Asthma (AZ-muh) is an inflammation and narrowing of the bronchial tubes, or air passages, in the throat. The muscles in the airway constrict, or tighten, and extra mucus is produced, blocking the normal flow of air into the body. During an asthma "flare," or attack, a person may experience wheezing, coughing, tightness of the chest, trouble breathing, and chest pain. In some cases, asthma can be life threatening.

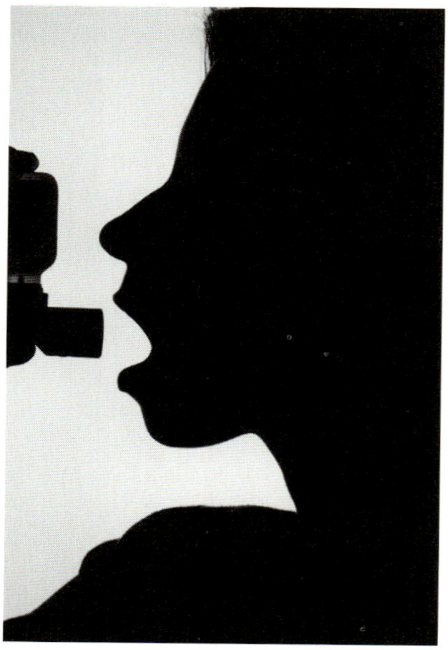

A person with asthma uses an inhaler to treat the symptoms of the disease. A puff of medication inhaled into the lungs makes breathing easier and relieves other asthmatic symptoms.

Asthma is becoming a severe problem across the nation and around the world. Between 1980 and 1994, the incidence of asthma increased by 75 percent in the United States. The CDC reports that in 1999, 26.7 million Americans were diagnosed as having asthma at some time in their lives.

No one knows exactly what causes asthma. Researchers believe that this condition develops early in life as a result of genetic and environmental causes. Medical professionals believe that asthma can be inherited because it seems to run in some families. However, doctors are certain that some environmental conditions can exacerbate, or worsen, asthma.

Certain triggers, such as exercise, cold weather, viruses, and stress, can set off an asthma flare. Asthma can also be triggered by contaminants, including air pollution, cigarette smoke, pet dander, chemicals, dust, pollen, perfumes, and mold and mildew. As a result, people who live in urban or industrial areas and who smoke or live with someone who smokes have a higher risk of developing asthma.

Minorities and people living in low-income areas are more at risk from asthma than are other populations. According to the CDC, due to asthma, African Americans have hospitalization and death rates that are three times higher than those of white Americans. Researchers have identified other risk factors that increase the chances of getting asthma, including having a parent with asthma, low birth weight, respiratory infections, and obesity.

TREATING ASTHMA

Asthma is a serious and sometimes deadly condition. According to the National Institute for Allergy and Immune Disorders (NIAID), asthma causes the deaths of 5,000 Americans each year. Asthma cannot be cured. However, it can be controlled and treated.

The chance of having an asthma flare can be reduced or, in some cases, prevented. People with asthma must learn to manage their condition and minimize any serious consequences of this disease. Asthmatics should learn to recognize the warning signs that signal a flare. These signs include shortness of breath, chest pain, coughing, and wheezing. If possible, people suffering from asthma should avoid the triggers that cause asthma flares.

As researchers learn more about asthma, they have developed better medications to treat asthma flares, including inhalers, pills, and immunotherapy shots. People suffering from asthma flares should seek emergency medical treatment immediately.

Asthma Facts

- Each year, 500,000 Americans are hospitalized because of asthma.
- More females than males have asthma.
- Children are more likely to get asthma than adults.
- Among children, asthma is the most common chronic illness.
- Asthma is the number one cause of hospitalization of children under the age of fifteen.
- Childhood asthma accounts for 10 million lost school days each year.
- Between 200,000 and 1 million asthmatic children have their asthma worsened by exposure to secondhand smoke.

SINUSITIS

Sinusitis (sye-nuh-SYE-tiss) is any infection or inflammation of the sinuses. The sinuses are air-filled chambers in the head and the cheekbones that drain into the nasal cavities. Symptoms of sinusitis include headache, aching of the teeth and jaws, stuffy nose, green or yellow mucus from the sinuses, swelling of the eyes, fever, and fatigue. In rare cases, sinusitis can lead to brain infection and other conditions.

Sinusitis can be caused by anything that inflames and irritates the sinuses. It often begins when a cold virus inflames the sinuses. If the cold doesn't go away after a week or two, chances are that it has developed into sinusitis. Doctors diagnose the condition by performing physical examinations. They may also take X-rays or perform other imaging tests. A person with sinusitis may be treated with antibiotics, pain relievers, and decongestants. If sinusitis recurs, surgery may be suggested. This type of surgery widens the openings that allow the sinuses to drain into the nasal passages.

Each year, about 37 million Americans are affected by sinusitis. People with allergies, asthma, and weakened immune systems are at a higher risk of developing sinusitis. The chances of getting sinusitis can be reduced by avoiding cigarette smoke, air pollution, alcohol, and chlorine. These substances all irritate and inflame the nose and sinuses.

Dr. Richard Hurt of the Mayo Clinic looks at a dramatic representation of the effects of secondhand smoke on children whose parents are smokers.

SMOKING-RELATED RESPIRATORY PROBLEMS

Did you know that smoking is the number one cause of preventable death in the United States? It puts people at risk for a variety of respiratory and other conditions, including colds, flu, bronchitis, and pneumonia. Smoking affects other body systems, as well.

Smoking is a big problem in the United States. According to the CDC, 3,000 children become regular smokers each and every day. The CDC also reports that one-third of these children will die what it calls a "smoker's death": from cancer, heart disease, or lung disease.

Those who smoke are putting others at risk, as well as themselves. Secondhand smoke—the smoke and chemicals that nonsmokers ingest from other people's tobacco smoke—can be deadly. Secondhand smoke contains over 4,000 different substances, at least forty of which have been shown to cause cancer in animals or humans. The ALA says that, each year, about 3,000 people die of lung cancer caused by secondhand smoke.

Each year, as many as 300,000 children under the age of eighteen months who breathe in secondhand smoke because their parents smoke become ill with a respiratory infection, such as pneumonia or bronchitis. Children of smokers are also more likely to suffer from coughing, excess mucus production, and fluid buildup in the inner ear than children whose parents don't smoke. In addition, secondhand smoke worsens cases of asthma, especially in children, irritating the eyes, nose, throat, and lungs.

Many health problems have been directly linked to cigarettes and other tobacco products. Here are just a few of them.

LUNG CANCER

Lung cancer is a cancer of any part of the lungs. A *cancer* is one of more than 100 different diseases in which a group of abnormal, or cancerous, cells divide without control in the body. A group of cancer cells is called a *malignant tumor.*

Cancers can spread from one part of the body to another when cancerous cells break away from the main tumor and enter the blood or lymphatic (lim-FAT-ik) streams. These cancerous cells are then transported to other sites in the body.

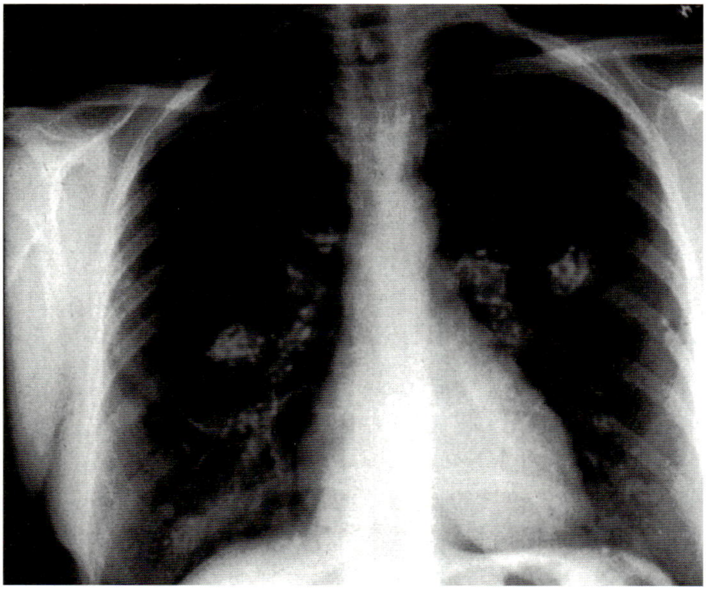

This X-ray shows a cancerous lung.

Lung cancer is the most common type of cancer among both men and women. According to the American Cancer Society (ACS), nearly 157,000 Americans died in 2000 because of lung cancer. Although the rate of lung cancer among men is decreasing, it is increasing among women.

Primary lung cancer begins when abnormal cells begin to grow in the lungs, triggered by tobacco smoke, chemicals, or other substances. Most lung cancers begin in the lining of the bronchi and develop over a number of years. As a result, most people diagnosed with this condition are more than sixty years old. Symptoms include a persistent or bloody cough, hoarseness, headache, fever, wheezing, weight loss, and chest pain.

Secondary lung cancer occurs when a cancer from another part of the body spreads to the lungs. This type of cancer is also known as *metastatic* (met-uh-STAT-ik) lung cancer. Some types of cancer that commonly travel to the lungs include breast, colon, esophageal, and bone cancers.

Lung cancers can be separated into two main groups, based on what the cancerous cells look like under a microscope. These two groups are non-small cell lung cancer and small cell lung cancer.

Non–small cell lung cancer (NSCLC) is the more common type of lung cancer. Most kinds of NSCLC begin in the lining of the lungs. This cancer usually grows and spreads slowly.

Small cell lung cancer (SCLC), also known as oat-cell cancer, is much less common than NSCLC. However, it spreads more quickly and is more likely to spread to other organs in the body. This type of lung cancer is almost always caused by smoking.

CANCER CAUSERS

The tendency to get cancer runs in families. This may be the result of inherited genes, similar lifestyles, or environmental factors that the family shares. There are some factors that can greatly increase a person's risk of getting cancer. Smoking is one such factor. Today, smoking is the major cause of lung cancer. About 87 percent of all lung-cancer cases can be directly related to smoking. It also causes cancers of the larynx and esophagus. Smokers who are exposed to environmental toxins have an even greater chance of getting lung cancer.

Exposure to radon is considered the second most common cause of lung cancer in the United States. *Radon* is a naturally occurring gas that is found in soil and rocks. Radon may leak into a home through cracked foundations, faulty insulation, or pipes, drains, and walls. Twelve percent of all lung-cancer deaths are caused by radon exposure.

Other chemicals that cause cancer include asbestos, uranium, arsenic, and some petroleum products. Many people are exposed to these chemicals in the workplace. These air pollutants have been shown to damage lung cells and have been linked to lung cancer.

Air pollution also has been linked to lung cancer. A recent study revealed that people who are exposed to long-term pollution from cars, factories, and power plants face a higher risk of the cancer. The greater the pollution and the longer the exposure, the higher the risk of lung cancer and other health problems.

DIAGNOSING AND TREATING LUNG CANCER

To diagnose lung cancer, doctors may take an X-ray or a *computerized tomography* (tuh-MAH-gruh-fee) scan—also known as a CT scan—of the chest. A CT scan uses a thin beam of electromagnetic radiation to create a three-dimensional picture of what's inside the body. These tests help doctors find abnormal spots in the lungs that may be cancerous tumors. Another method of detecting the disease is *sputum cytology* (sye-TAHL-oh-jee), in which doctors examine the contents of mucus from within the lungs. Doctors can also perform a *biopsy,* taking a sample of lung tissue to examine under a microscope. They use a needle or a small tube called a *bronchoscope* (BRON-koh-scope) to extract the tissue. The tissue may also be taken out surgically.

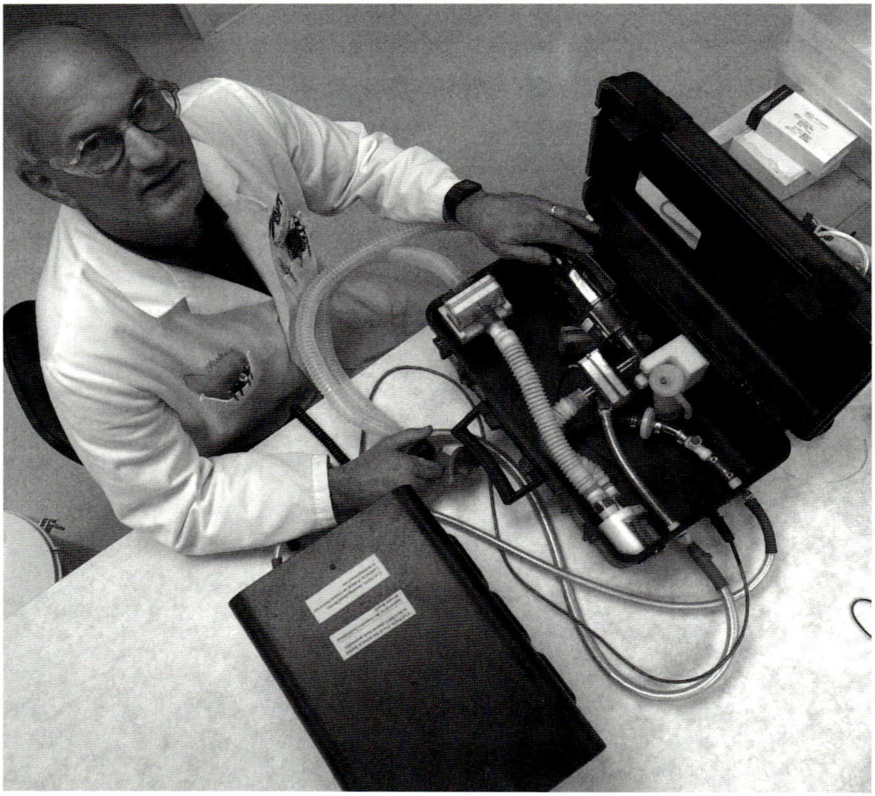

A device known as an inhalator administers medication to people with lung cancer.

Once lung cancer has been diagnosed, doctors need to know how much the cancer has grown or spread. They may use MRIs, CT scans, or other types of scanning tests to look inside the body and see what is going on. How the cancer is treated will depend upon the type of cancer, the cancer's size and location, and the patient's health.

Surgery could be performed to remove the cancerous tumor or tumors from the body. In some cases, surgeons remove parts of the lung, including the lobes. When a lobe is removed, the procedure is called a *lobectomy* (loh-BEK-tuh-mee). Serious cases may require the removal of an entire lung. This procedure is called *pneumonectomy* (noo-moh-NEK-tuh-mee). Some cancers, however, cannot be operated on because of their size or location.

Some cancer cases call for chemotherapy. Most chemotherapy drugs are given intravenously, although some are designed to be taken orally. Other cancers are treated with radiation. Patients undergo doses of radiation aimed at the cancerous area. The radiation shrinks cancer cells, increasing the chances that other forms of treatment will be effective. It can also ease the pain of someone suffering from lung cancer.

Photodynamic therapy is a new therapy that uses lasers to kill cancerous cells. The patient is first injected with a special chemical dye. Cancer cells absorb and hold the dye longer than normal cells do. Doctors can then target the dye-stained, cancerous cells for destruction.

Although there have been great strides in detecting and treating lung cancer, doctors say that prevention is still the best medicine. They recommend that people quit smoking and limit their exposure to secondhand smoke and other airborne toxins.

Stages of Cancer

In order for doctors to know how best to treat a patient's cancer, they must know what stage the cancer is in. Stages are the way that doctors describe the severity of a cancer case. Stages may range from Stage 0 (least serious) to Stage V; each type of cancer is described by different stages. Here are the four stages of NSCLC.

- **STAGE I:** The cancer is found in the lung only and not in the tissues around the lung.
- **STAGE II:** The cancer has spread to nearby lymph nodes.
- **STAGE III:** The cancer has spread to structures near the lung, to the lymph nodes in the area that separates the two lungs, or to the lymph nodes on the other side of the chest or in the lower neck.
- **STAGE IIIA:** Surgery to remove the tumor may be an option.
- **STAGE IIIB:** Surgery to remove the tumor is usually not an option.
- **STAGE IV:** The cancer has spread to other parts of the body. Stage IV lung cancer is usually not curable.

CHRONIC BRONCHITIS

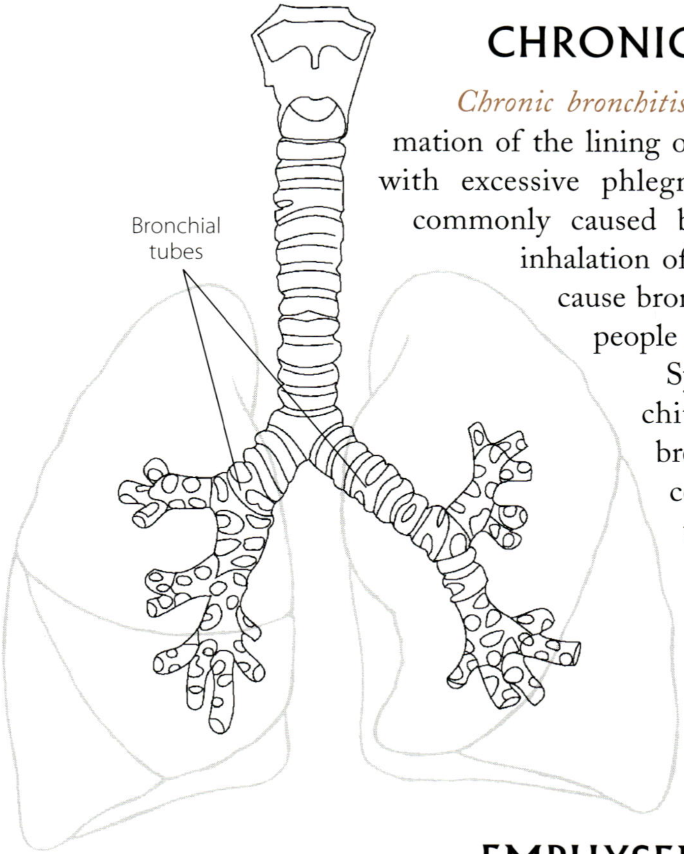

Chronic bronchitis (bron-KYE-tiss) is an inflammation of the lining of the bronchial tubes, along with excessive phlegm production. It is most commonly caused by cigarette smoking. The inhalation of certain chemicals may also cause bronchitis. More than 8 million people suffer from this condition.

Symptoms of chronic bronchitis include shortness of breath and a mucus-producing cough. Chronic bronchitis, unlike acute bronchitis, lasts for a long time—weeks and sometimes months. It may return again and again. Chronic bronchitis may lead to emphysema.

EMPHYSEMA

Emphysema (em-fih-SEE-muh) occurs when the walls between the air sacs of the lungs lose their ability to stretch. The walls become weakened and unable to deflate and inflate properly. When this happens, air becomes trapped inside the sacs and the exchange of oxygen and carbon dioxide is inhibited.

Emphysema is a progressive condition. As the disease progresses, the lungs become filled with more air, making it extremely difficult for a person to breathe. The lungs expand, and a person may have a barrel-chested appearance as a result.

The symptoms of this serious and debilitating disease include coughing, shortness of breath, wheezing, and difficulty breathing. The exact cause of emphysema is unknown, but medical professionals say that smoking and air pollution are high-risk factors. In addition, people with bronchitis and asthma face a greater chance of getting this condition. Nearly 3 million Americans suffer from emphysema.

One experimental treatment that is being tried on patients with emphysema is called *lung reduction.* Lung reduction is a surgical procedure to remove damaged sections of a lung. Doctors hope that this will give the healthy sections of the lung more room to expand.

DISEASES CAUSED BY MICROORGANISMS

Many respiratory disorders are caused by viruses, bacteria, and other organisms. These infection-causing organisms can create health problems throughout the respiratory system. They can attack and affect the lungs, air passages, nose, and mouth.

PNEUMONIA

Pneumonia (noo-MOHN-yuh) is a common respiratory infection. Each year, as many as 4 million Americans suffer from pneumonia. This infection of the lower respiratory tract affects the small bronchi and air sacs of the lungs. Pneumonia can be caused by bacteria, viruses, or fungi. There are more than fifty different types of pneumonia.

People of all ages can get pneumonia, but older people and people with weakened immune systems are especially at risk. For them, pneumonia is a serious, potentially deadly condition. It is the fifth leading cause of death in the United States. In 2000, more than 64,800 Americans died as a result of pneumonia.

> **Fast Fact**
>
> "Double pneumonia" is the popular term for a pneumonia that has infected both lungs.

Cases of pneumonia caused by bacteria are usually more serious than those with other causes. More than half of all cases of pneumonia are caused by bacteria known as *Streptococcus pneumoniae* (strep-toh-KAHK-uss noo-MOHN-yay), also called pneumococci (noo-moh-KAHK-sye). Pneumonia caused by this organism, known as *pneumococcal* (noo-moh-KAHK-ul) *pneumonia,* usually affects children under the age of two and adults over the age of sixty-five. Symptoms include high fever, chills, coughing, chest pains, and shortness of breath. In addition, patients may experience headaches, muscle aches and pains, nausea, and fatigue.

Another type of pneumonia is *Legionnaires' disease.* This rare respiratory infection was first diagnosed in 1976 when more than 220 members of the American Legion became ill during a convention at a Pennsylvania hotel. It was later believed that this bacterial pneumonia was spread through contaminated water used in the hotel's air-conditioning system. Thirty-four people died.

Doctors diagnose pneumonia by performing physical examinations. They may also use chest X-rays and blood tests to look for the presence of organisms known to cause the condition. Different types of pneumonia are treated in different ways. For example, patients with bacterial pneumonia are treated with antibiotics. Viral pneumonias, however, will not respond to antibiotics. Instead, doctors recommend plenty of rest and lots of fluids.

The best way to prevent some types of pneumonia is to be vaccinated against the disease-carrying bacteria. All people over the age of sixty-five should be given the vaccine, as should people with other conditions that place them at risk for pneumonia. Because pneumonia is contagious, health experts recommend washing the hands frequently and staying away from people who are sick with flulike symptoms.

Super Germs on the Loose

Antibiotics can work wonders fighting such bacterial infections as pneumococcal pneumonia. It is important that people with this disease take their medicine as directed and finish all of it. When people don't finish their antibiotics, the pneumonia bacteria may become resistant to future antibiotic treatments. This means that medicine won't work the next time pneumonia strikes.

The problem of drug-resistant bacteria, sometimes called "super germs," is on the rise. In 2000, 34 percent of all pneumonia-causing bacteria were resistant to antibiotics. Doctors are trying to fight this disturbing new trend by relying on other treatments for many common bacterial infections.

VIRAL INFECTIONS

The two most common types of *viral upper respiratory infections* are colds and the flu. Other types of upper respiratory infections include *laryngitis* (laar-in-JYE-tiss), or inflammation of the larynx, and *pharyngitis* (faar-in-JYE-tiss), or inflammation of the pharynx.

A *cold* is a minor, contagious illness that affects the throat and nose. Symptoms include nasal congestion, runny nose, sneezing, sore throat, and headache. These symptoms usually disappear within one to two weeks.

Researchers have identified more than 200 different cold viruses. According to the National Center for Health Statistics (NCHS), 62 million Americans suffer from colds each year. There is no cure, but patients can help themselves feel better with medicine, rest, and plenty of liquids.

Influenza (in-floo-EN-zuh), also know as the flu, is another contagious—and common—viral infection. The flu is spread through sneezing and coughing. It can be a very serious condition for some people: Each year, more than 100,000 Americans are hospitalized and about 20,000 die as a result of the flu.

There are three main types of flu: A, B, and C. Type A is the most serious. This type of flu is responsible for the yearly flu epidemics around the United States. During some epidemics, as many as 50 million Americans may be infected. Type B is less common than Type A, and Type C is the rarest of all. Type C usually causes only mild symptoms.

Health and History: The Flu Epidemic of 1918

Today, when people think of the flu, they think of an uncomfortable—but curable—illness. In September 1918, however, the story was quite different. A flu epidemic had spread across the United States. About one in every four Americans suffered from the sickness. More than 600,000 Americans died as a result of the disease. While many of the fatalities were the very old and the very young, many others were between the ages of twenty and twenty-nine. It was the worst epidemic ever to hit the United States.

American soldiers, arriving in Europe to fight in World War I (1914–1918), unknowingly brought the virus with them. From Europe, the killer organism spread to Russia and Africa, and then on to Asia. By the time that the epidemic finally ended, the international death toll was staggering. Historians estimate that between 20 million and 40 million people died during the epidemic—far more than the 15 million soldiers who died during World War I. Today, it is still not known what made this strain of flu virus so deadly.

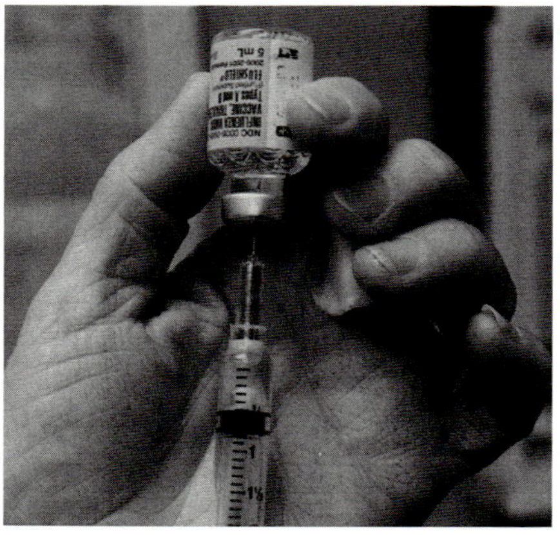

An inoculation with the vaccine for influenza A and B can help control the flu. Left untreated, the flu can develop into more serious conditions, such as pneumonia, in people in certain vulnerable groups.

Symptoms of the flu include high fever, headache, aching joints and muscles, coughing, difficulty breathing, and sore throat. The condition is usually treated with rest, fluids, and over-the-counter medicines to ease aches and pains. Most symptoms disappear within ten days.

People can fight the flu before it hits by getting an annual flu shot. Older people and other vulnerable groups should always get an annual immunization. The shot's effectiveness varies, however. Those who receive the shot may still become ill with the flu.

CROUP

Croup is a viral infection of the upper air passages. Tissues in the respiratory tract swell so less air is able to get to the lungs. Croup usually occurs in children under the age of three whose air passages haven't yet widened enough. They most often become ill during the fall and winter months.

Symptoms of croup include a barking cough, difficulty breathing, fever, and cyanosis, a bluish tinge of the skin. Most cases can be treated with rest, fluids, and exposure to moist air. Other more serious cases may require hospitalization.

PLEURISY

Pleurisy, also known as pleuritis, is an inflammation of the pleura around a lung. Most cases of pleurisy are caused by viral infections, but it may also result from other health conditions including pneumonia, tuberculosis, or a pulmonary embolism.

Symptoms of pleurisy include chest pain, shortness of breath, cough, and fever. In many cases, fluid collects at the site of the inflammation. When this occurs, chest pains may ease, even though the condition is worsening.

Doctors may remove fluid from the lungs in a procedure called a *thoracentesis.* During this procedure, fluid is drawn from the lung with a needle and a syringe. Doctors can then examine the fluid for clues about what caused the infection. If the infection was caused by bacteria, doctors treat it with antibiotics. Pleurisy, as a result of a viral infection, must be left to run its course.

ACUTE BRONCHITIS

This condition is almost always caused by viruses that affect the bronchial lining. In most cases, the viruses are the same ones that cause the common cold. Symptoms of acute bronchitis include a mucus-producing cough and wheezing. Acute bronchitis usually goes away within ten days.

BACTERIAL INFECTIONS

Diphtheria (dif-THEER-ee-uh) is caused by the bacterium *Corynebacterium diphtheriae* (koh-rye-nee-bak-TEER-ee-um dif-THEER-ee-ay). It most commonly affects the nose, throat, tonsils, and larynx, but it may also affect the skin and cause damage to the circulatory and nervous systems. The bacteria attack and destroy tissues in the affected area.

Diphtheria, found only in humans, is spread when an infected person coughs or sneezes. Symptoms include sore throat, fever, barking cough, and watery or bloody nasal discharge. In some cases, the disease's first symptoms are skin lesions.

Vaccines are available to prevent people from getting diphtheria. The vaccine is given to children, and then booster shots are required every ten years. As a result, diphtheria is rare in many parts of the world. However, people living in crowded conditions who do not receive immunization shots are at high risk for this disease.

Pertussis, also known as whooping cough, is a contagious bacterial disease. Although pertussis was once a disease most commonly associated with children, people of any age can be affected by it. Symptoms of this upper respiratory infection include severe coughing that ends in a whooping noise. Other symptoms may be similar to cold symptoms, including fever and a stuffy nose. Illness can last for up to six weeks.

Doctors treat pertussis with antibiotics to kill bacteria. If left untreated, pertussis may result in death due to a lack of oxygen to the brain. A vaccine exists to prevent people from suffering from pertussis, and health officials recommend that all people receive the vaccine.

TUBERCULOSIS

Tuberculosis (too-ber-kyoo-LOH-siss), or TB, is an infectious bacterial condition that mainly affects the lungs. The disease can be spread when an infected person coughs or sneezes. TB is sometimes, although rarely, contracted through the skin. For example, people risk becoming infected with TB if they get a tattoo with unsterilized, or dirty, equipment.

The bacterium that causes TB spreads quickly in crowded conditions. In the past, TB was the most serious health threat in the United States. In the 1940s, the first antibiotics to treat the disease were discovered, and the number of TB cases began to fall. In 1985, however, TB began making a comeback in the United States. Between 1985 and 1992, the number of TB cases rose steadily. Although that trend has since been reversed, thousands of Americans still suffer from the disease.

> **Fast Fact**
>
> The WHO has named tuberculosis as one of the "big three" infectious killer diseases around the world. The other two are AIDS and malaria.

TB continues to be a very real health problem all around the world. According to the NIAID, one-third of the world's population is infected with the bacteria that cause TB. Three million of these people die annually. Not all people infected with TB will become ill, however. In fact, only one in ten people with the bacteria will ever develop the disease. Those who do not develop the disease are said to have *latent TB.* People with latent TB will never show any symptoms of the disease. The TB bacteria will never become active and may even die after many years.

TB can be a problem when people's immune systems become weak and are less able to fight the bacteria. The elderly, people infected with the HIV virus (which causes the disease AIDS), and people with other health conditions may have weakened immune systems. TB is also a problem among particular populations of people, including the homeless, prisoners, and people addicted to drugs and alcohol.

During the early stages of TB infection, people may not even know they have a disease. As the TB bacteria multiply, however, symptoms begin to appear—coughing, chest pain, shortness of breath, fever, chills, and fatigue. Later, as TB bacteria begin to destroy the tissue in the lungs, patients may begin coughing up blood and phlegm. TB can also cause damage to the kidneys, bones, lymph nodes, and brain. If the condition is not treated quickly, it can be fatal.

Doctors diagnose TB with a skin test and chest X-ray. If a patient tests positive for TB, doctors use antibiotics to treat the disease. These antibiotics must sometimes be taken for up to a year. It is very important for patients to complete all their medication or they can develop drug-resistant TB. This type of TB is extremely difficult to treat and often fatal.

Researchers are trying to put an end to TB. They are working to develop vaccines that will prevent people from ever getting the condition.

Frédéric Chopin

Health and History: TB Through the Years

Tuberculosis has existed for centuries. Traces of the disease have been found in Egyptian mummies from 2400 B.C.E. In the past, it had been called consumption, the English disease, and the white plague. Throughout history, hundreds of doctors have tried to discover its causes and cures. Some of the stranger treatments for the disease included drinking elephant blood, bathing in urine, and living underground.

TB has claimed thousands of lives through the years, from the rich and famous to the poor and unknown. Famous victims include poet John Keats (1795–1821), composer Frédéric Chopin (1810–1849), novelists Robert Louis Stevenson (1850–1894) and Charlotte Brontë (1816–1855), King Edward VI of England (1537–1553), first lady Eleanor Roosevelt (1884–1962), and gunslinger Doc Holliday (1851–1887).

In 1882, German doctor Robert Koch (1843–1910) discovered that TB was caused by bacteria. It wasn't until 1944, however, that an antibiotic to successfully fight the disease was developed. Since then, researchers have developed a number of drugs that work together to cure the disease.

ANTHRAX

Anthrax is a disease caused by the bacterium *Bacillus anthracis* (buh-SILL-us an-THRAY-siss). The bacteria that cause anthrax can be inhaled, be ingested in contaminated food, or enter the body through breaks in the skin. Anthrax cannot be spread from person to person.

Fast Fact

In 2001, anthrax spores sent through the U.S. mail were responsible for the deaths of a number of people, including two Washington, D.C., postal workers.

There are three serious types of anthrax that affect humans: inhalation, cutaneous, and intestinal. People become infected with *inhalation anthrax* when they inhale anthrax spores, or cells. Inhalation anthrax is often fatal. During the first stage of infection, symptoms are similar to those of the flu or a cold: low-grade fever, headache, fatigue, sweats, and chest discomfort. Within one to five days, the patient may suffer a high fever, serious breathing problems, and shock. Ninety percent of all patients who enter the second stage of anthrax infection die.

People become infected with *cutaneous* (kyoo-TAY-nee-uss) *anthrax* when the bacteria enter wounds or breaks in the skin. This type of anthrax is the most common and is usually caused by contact with infected livestock. It causes painful blisters and skin lesions. Lymph nodes near the site of the infection may also become infected. The lesions eventually dry up and disappear on their own, but it is important to get this type of anthrax treated. In two out of ten cases, the infection can spread to the blood and cause illness and death.

Workers clad in protective suits prepare to decontaminate the Hart Senate Office Building after anthrax spores were discovered there in 2001.

People become infected with *intestinal anthrax* by eating contaminated food. Symptoms of this very rare infection include nausea, loss of appetite, vomiting, and fever. More serious symptoms may include diarrhea, abdominal pain, and vomiting of blood. This type of anthrax has never been diagnosed in the United States.

Doctors treat anthrax with antibiotics to prevent infection. Although there is currently a vaccine for inhalation anthrax, it is not recommended or available to the American public. Because anthrax can be used as a weapon, however, the vaccine is required for all United States military personnel. The use of bacteria and other biological agents in this manner is known as bioterrorism. Experts say that, although anthrax has been used in recent years to terrorize Americans, it probably could not be used as a weapon on a large scale.

PNEUMONIC PLAGUE

Pneumonic plague is a contagious bacterial infection that can be deadly. Humans contract the plague from fleas that have bitten infected rodents. The disease is then spread from one person to another by face-to-face contact. In the United States, pneumonic plague most commonly affects residents of the West and Southwest.

Pneumonic plague attacks the lungs, causing difficulty breathing, fever, chills, headache, cough, and chest pain. In the past, the plague was a death sentence. Today, however, the plague can be cured with antibiotics. Patients with this disease must be isolated to avoid spreading it to others. They may be hospitalized to receive oxygen and other respiratory treatments. With quick treatment, 95 percent of those who contract this disease can now be cured.

Experts on bioterrorism warn that the bacteria that cause pneumonic plague could be used as a weapon. The bacteria could be dispensed among an unsuspecting public by aerosol sprays. Many medical communities have begun performing drills to prepare them for such an attack.

FUNGAL AND OTHER TYPES OF INFECTION

Some types of respiratory infections are caused by fungi. People develop such infections after breathing in the spores of certain types of fungi. As a result, the lungs are the organs most commonly affected by fungal infections. Although many people can come in contact with these spores without serious problems, people with lung conditions and weakened immune systems may develop respiratory infections.

Symptoms of a fungal respiratory infection may include shortness of breath; coughing up blood, mucus, or pus; chest pain; fever; and weight loss. In severe cases, fungal infections may result in scarred lung tissue. Fungal infections cannot be spread from one person to another. Doctors often treat these types of infections with medications that kill the fungi. With some types of infection, surgery may be required to remove the fungi.

OCCUPATIONAL LUNG DISEASE

Occupational lung disease is any disease of the lungs that is caused by long-term exposure to irritating or toxic substances in the workplace. These diseases include lung cancers, asthma, and other respiratory illnesses.

Occupational lung diseases are highly preventable. Proper workplace ventilation, protective equipment, and safe working procedures can all help to reduce the chances that a worker will be affected with job-related respiratory ailments. Smokers are at higher risk for severe occupational lung disease.

ASBESTOSIS

Asbestosis (ass-bess-TOH-siss) causes scarring to the lung tissues. Scarred lungs do not expand and contract properly. Workers who breathe in *asbestos,* any of several fibrous, heat-resistant minerals, are affected. Asbestos is found naturally in rocks, and before 1989, it was used in building materials, paper, and automotive products. Plumbers, pipe fitters, carpenters, mechanics, welders, and miners are more likely to come in contact with the material than others. In addition, if a worker has asbestos particles on his or her clothing, the entire family may be affected.

Asbestosis is not the only respiratory condition caused by exposure to asbestos. Other serious conditions include mesothelioma (mez-uh-thee-lee-OH-muh), or cancer of the pleura, and pleural effusion, which is fluid buildup in the pleura.

SILICOSIS

Every year, 300 Americans die from *silicosis* (sil-ih-KOH-siss). This condition is caused by inhalation of silica dust, which is found in sand, rock, and minerals. Those most often affected by silicosis are quarry workers, sandblasters, stonemasons, and glassworkers. People in these industries can reduce the risk of silicosis by wearing protective equipment and not smoking. Smoking increases the chances of lung damage from this disease.

There are two types of silicosis, acute and chronic. *Acute silicosis* results from heavy exposure to silica dust over a period of months.

Chronic silicosis occurs over years of exposure to the dust. Both types of silicosis result in inflammation and scarring of the lungs, as well as the development of growths or *nodules* on the lungs.

> **Fast Fact**
>
> According to the WHO, silicosis is the world's oldest occupational disease. It has been affecting people since ancient Greek times.

Silicosis is a serious condition. There is no way to reverse the damage done to the lungs by this disease. Silica inhalation has been shown to lead to other health conditions, as well, including bronchitis and pneumonia. It may also be linked to lung cancer.

PNEUMOCONIOSIS

Also known as black lung disease, *pneumoconiosis* (noo-moh-koh-nee-OH-siss) is often contracted by coal miners who have inhaled coal dust for many years. Dust particles stick to lung tissue, causing inflammation. Eventually, scar tissue begins to form and destroy lung cells.

J. Davitt McAteer (right), the former assistant secretary of labor for mine safety and health, talks with a mine worker about the dangers of coal dust. The Department of Labor is studying black lung disease, which is caused by coal dust.

Symptoms of pneumoconiosis include shortness of breath and coughing. The severity of the disease depends upon the length and extent of exposure to the coal dust. Serious cases may result in disability. There is no way to treat this disorder.

BYSSINOSIS

More commonly called brown lung disease, *byssinosis* (bih-zih-NOH-siss) affects textile workers. Sometimes known as "cotton worker's lung," this condition is caused by inhaling cotton, hemp, and flax dust. The dust obstructs the small airways, causing chest tightness, coughing, and wheezing. If workers with byssinosis continue to be exposed to the textile dust, they may develop permanent lung damage.

Acute Mountain Sickness

Also known as altitude sickness, *acute mountain sickness* results when a person ascends too rapidly from a lower to a higher altitude. People who normally suffer from this condition include skiers, mountain climbers, and some airplane passengers. All airplane cabins are carefully pressurized, however, so the incidence of mountain sickness on air flights is marginal.

Although most cases of acute mountain sickness are not serious, the condition can cause illness and even death. Fluid can build up in the lungs, causing a lack of oxygen throughout the body. Symptoms of mild mountain sickness include headache, fatigue, nausea, and dizziness. Symptoms of severe mountain sickness include shortness of breath, confusion, and coma, as the brain is deprived of the oxygen it needs.

7

THE IMMUNE SYSTEM

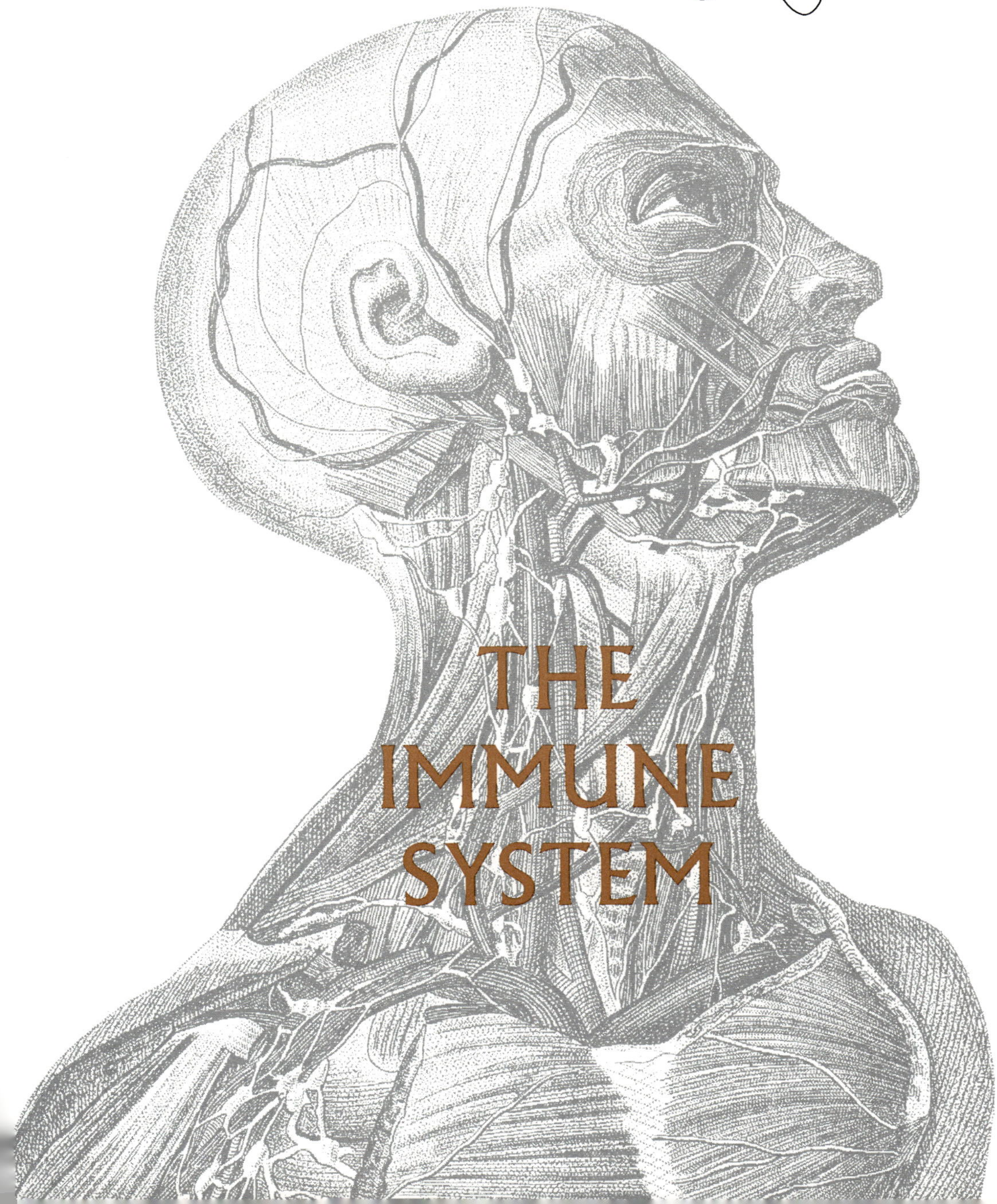

Every second of every day, people breathe in an enormous quantity of bacteria, viruses, and other organic substances. While some of these substances are harmless or even beneficial to the body, many are not. It is the job of the immune system to defend the body from harmful diseases and other problems. Without this system, the human body would be at the mercy of every bacterium, virus, and parasite that entered the body.

The immune system is made up of organs, cells, and special circulatory vessels that perform the job of fighting foreign substances in the body. It recognizes and attacks these foreign invaders as harmful. The immune system also attacks the body's own unhealthy or diseased cells.

A person's first line of defense is called *innate immunity.* Innate immunity includes the skin, the body's largest organ. The skin acts as a protective shield for the body. Other initial defense mechanisms include tears, mucus, and saliva. These substances contain enzymes that kill off bacteria, viruses, and other toxic organisms.

If a virus or other harmful organism gets past the first line of defense, the body responds with *adaptive,* or acquired, immunity. With adaptive immunity, the body learns ways to handle invading organisms.

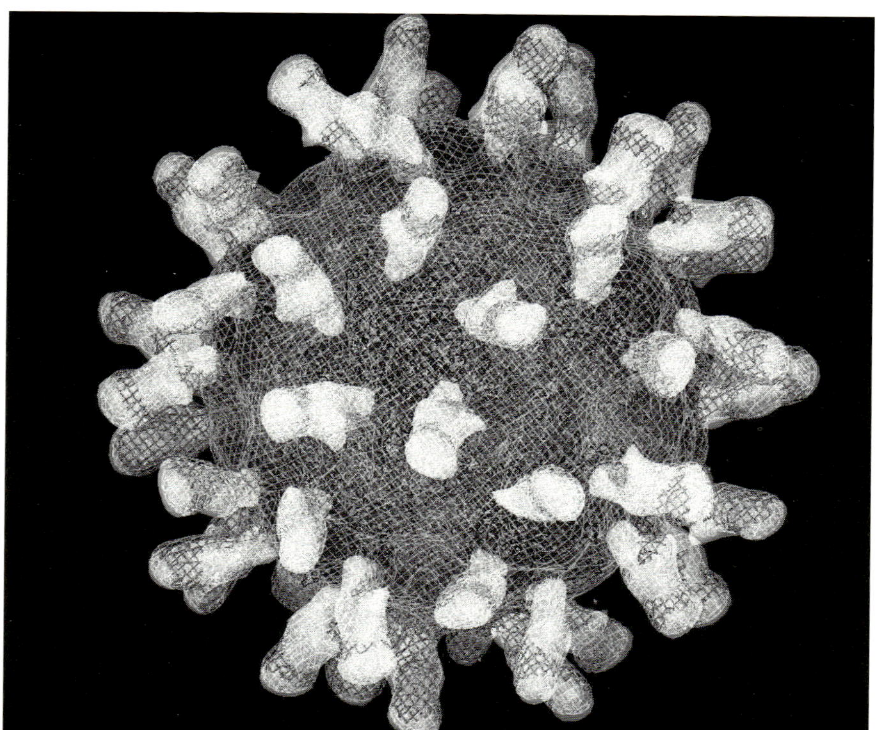

This photograph taken with an electron microscope shows the virus that causes the common cold.

The immune system includes the following parts:

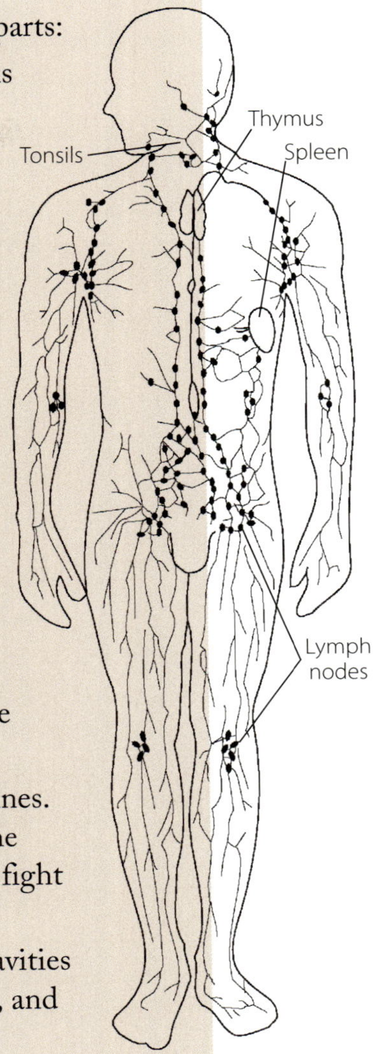

- The *lymph nodes* are small, bean-shaped organs found throughout the body. Infection-fighting cells are produced here.
- *Lymph* is the plasmalike fluid that carries infection-fighting cells throughout the body. Lymph is filtered from blood and eventually returns to blood. The *lymphatic vessels* are the tubes that lymph travels through. Lymphatic vessels are very similar to veins and capillaries.
- Located on either side of the throat, the *tonsils* fight foreign matter that has been inhaled or ingested.
- The *spleen,* located in the abdominal cavity, is the largest of the immune system organs. The spleen filters disease-causing organisms out of the bloodstream. Certain types of white blood cells are stored here.
- The *appendix* is connected to the large intestines. Scientists are not sure what the function of the appendix is, but they believe that it may help fight infection in the body.
- *Bone marrow* is the soft tissue found in the cavities of some bones. Blood cells are produced here, and B cells, which are a type of white blood cells, mature here.
- The *thymus* (THYE-muss) plays a very important role in the immune system. Stem cells produced in the bone marrow develop into special infection-fighting cells called T cells here.
- The *stomach* acts as part of the immune system by destroying ingested organisms with acids and enzymes.
- The *liver* produces some types of infection-fighting proteins.

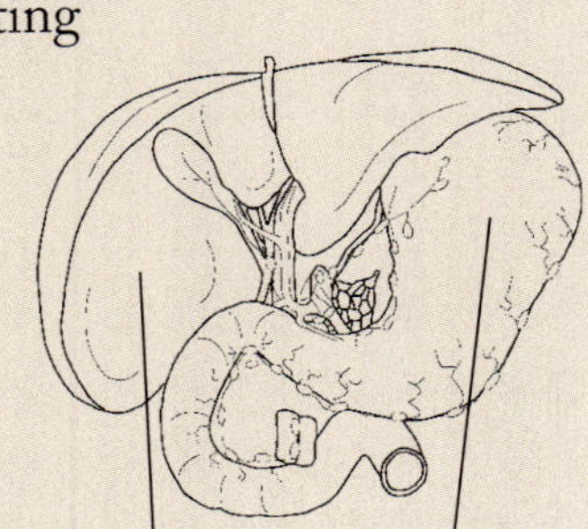

(reside in front of the spleen)

THE IMMUNE SYSTEM AT WORK

The immune system kicks into action when substances with antigens enter the body. Antigens are structures found on the surfaces of any foreign substances. The mission of the immune system is to attack and destroy all antigens.

When an antigen enters the body, an *inflammatory response* is the immune system's first reaction. During this type of response, tissues become inflamed. Special white blood cells called neutrophils arrive in the area and attempt to destroy the antigen.

Inflammatory response affecting several areas of the body is illustrated in this drawing.

If the inflammatory response isn't effective, an *immune response* is next. There are two types of immune response. One is an *antibody defense.* Special white blood cells, called *B cells,* begin to multiply. Some of the cells secrete antibodies, proteins that circulate through the body and attack antigens. The antibody defense destroys bacteria and toxins that attack from outside of cells. It can also prevent viruses from invading cells.

The other immune response is a *cellular defense. T cells,* another type of white blood cell, kill any leftover antigens and infected cells. The cellular defense kills antigens that harm the body by invading cells, including viruses.

For every antigen that enters the body, the immune system produces one unique type of antibody. Once it has attacked an antigen, the immune system will "remember" that substance. If the antigen shows up in the body again, the immune system will react even more quickly than before, in many cases actually preventing a person from becoming ill. This quick, effective reaction to an antigen is called *immunization.* A person can become immunized naturally by the immune system or artificially, through vaccination.

Symptoms of the immune system in action include pain, swollen lymph nodes, and fever. These symptoms show that the body is working hard to fight off foreign invaders.

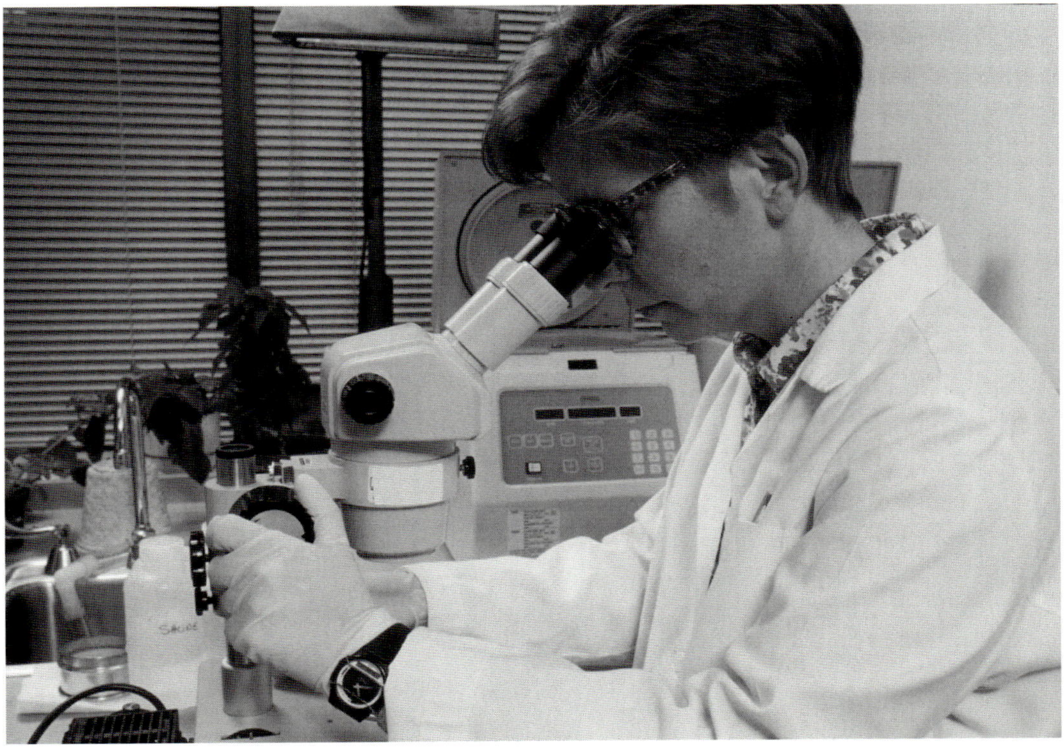

A researcher looks through a microscope to analyze blood components and determine the body's response to antigens.

WHITE BLOOD CELLS

White blood cells, also called *leukocytes* (LOO-koh-sites), play an important role in the immune system. There are several types of white blood cells, including lymphocytes, granulocytes, and phagocytes.

Lymphocytes (LIM-foh-sites) battle foreign organisms. Although they originate in the bone marrow as stem cells, some lymphocytes actually mature in the thymus. There are more than 1 trillion lymphocytes in the average human body. Lymphocytes are divided into several categories, including B cells, T cells, and natural killer cells.

B cells are made in the bone marrow. These lymphocytes produce different types of antibodies to fight different types of diseases. The antibodies produced by the B cell depend on the type of antigen that the B cell has identified. For example, the B cell makes one type of antibody to fight a cold virus and a different type to fight a flu virus. The antibodies circulate throughout the body in the circulatory and lymph systems, looking for substances that are foreign. When they find a substance that they have been made to identify, they attach to it, marking it to be destroyed.

T cells are lymphocytes that mature in the thymus. They attack and destroy foreign organisms. T cells circulate throughout the body in the blood and lymphatic vessels. Different types of T cells have different functions. *Helper T cells* activate B cells, spurring them to make more antibodies. By doing this, they can control the strength of the immune response. Helper T cells can also activate other T and disease-fighting cells. *Suppressor T cells* halt the creation of antibodies and other immune system responses. *Killer T cells* attack any leftover disease antigens, as well as any infected cells.

Natural killer (NK) *cells* destroy foreign cells. NK cells will attack anything that they identify as foreign, whether it is a cancer cell, a virus, or a bacterium.

> **Fast Fact**
>
> The yeast that most commonly causes infections in the human body is called *Candida.* This is not the same yeast that is used to make bread and wine.

Granulocytes (GRAN-yoo-loh-sites) are white blood cells that contain granules of chemicals and enzymes that fight foreign substances. Some types of granulocytes release histamine in response to an allergy. *Histamine* (HISS-tuh-meen) is a substance that causes the symptoms of allergic reactions. Other granulocytes contain chemicals that can harm parasites and other infectious organisms. Granulocytes are the most common type of white blood cell.

Phagocytes (FAGG-oh-sites) circulate through the body and engulf foreign substances. Some B cell antibodies are made to attract these types of cells. The antibodies coat the foreign substance, and the phagocytes then surround and destroy it.

Opposite:
This magnified photograph shows white blood cells above and red blood cells below.

Invading Organisms

What is the difference between bacteria and viruses? Between protozoa and fungi? Knowing the differences between these organisms and how they affect the human body allows doctors to treat diseases effectively.

- *Bacteria* are tiny, single-celled organisms that are present in the air, water, and earth. There are thousands of different types of bacteria. Some are helpful to humans. For example, some types of *Escherichia coli* (eh-shur-IK-ee-uh KOH-lye), more commonly known as *E. coli,* are normally found in healthy intestines. These strains of *E. coli* help the body to digest food. Other types of bacteria are not helpful at all. Examples of harmful bacteria include other strains of *E. coli,* as well as bacteria that cause the diseases strep throat, tetanus, and diphtheria.

These *E. coli* colonies were grown from samples taken from an outbreak of diseases caused by the bacteria in Washington state in 1996. They were used to identify the substrain of the bacteria and isolate the cause of the outbreak.

- *Viruses* are very simple organisms that are found everywhere. They are much smaller in size than bacteria. There are many different types of viruses, and they are all harmful to humans. Viruses are responsible for such diseases as AIDS, herpes, and flu.
- *Protozoa* are tiny, single-celled organisms. Protozoa eat bacteria. While most protozoa are not harmful, some types carry some very serious diseases, including malaria and giardiasis (jee-ar-DYE-uh-siss).
- *Fungi* are simple organisms that live by eating dead or dying tissue. Some types of fungi are helpful, such as yeast and the fungus from which we get the antibiotic *penicillin* (pen-uh-SILL-in). Other types of fungi are harmful. Fungi cause such diseases as thrush (a disorder of the mouth caused by overgrowth of a certain yeast) and meningitis (men-in-JYE-tiss).

8

IMMUNE SYSTEM DISORDERS

When the immune system doesn't work properly, people become ill more easily and take longer to recover. Sometimes, the immune system malfunctions and attacks its own healthy cells. Disorders caused by a malfunctioning immune system are called autoimmune disorders.

Autoimmune disorders can affect many different parts of the body. They may affect certain organs or entire body systems. One autoimmune disorder that affects the stomach lining is pernicious anemia. Rheumatoid arthritis is an autoimmune disorder that affects the entire musculoskeletal system.

CONGENITAL PROBLEMS

Some people are born with faulty immune systems. When this happens, their condition is called primary immune deficiency (PID). People with PIDs are at a high risk for infection and disease because their immune systems are abnormal. People with PIDs may get one infection after another. In addition, they have difficulty recovering from these infections.

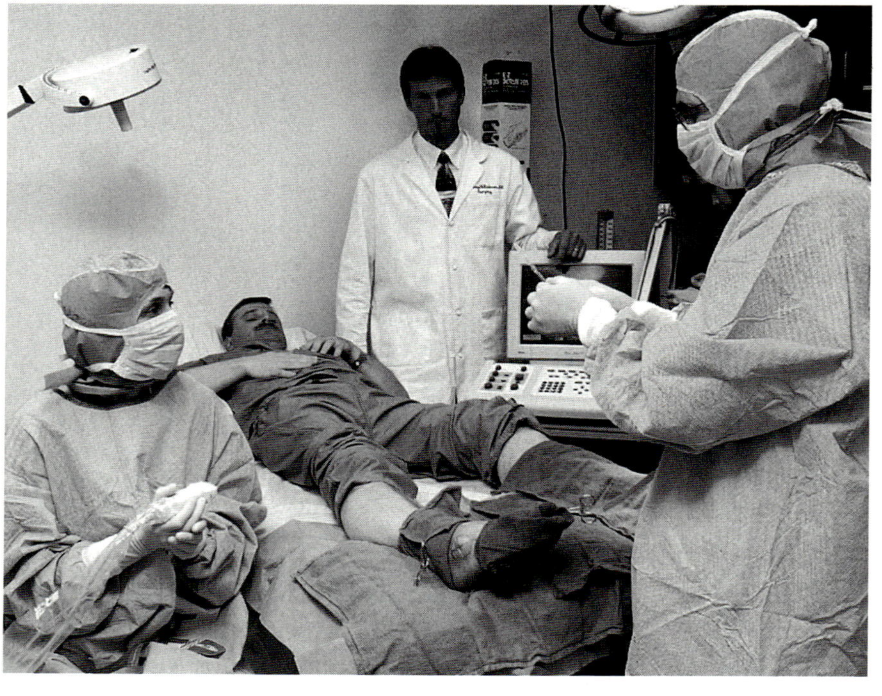

This patient is about to receive a gene-therapy injection to treat muscular dystrophy.

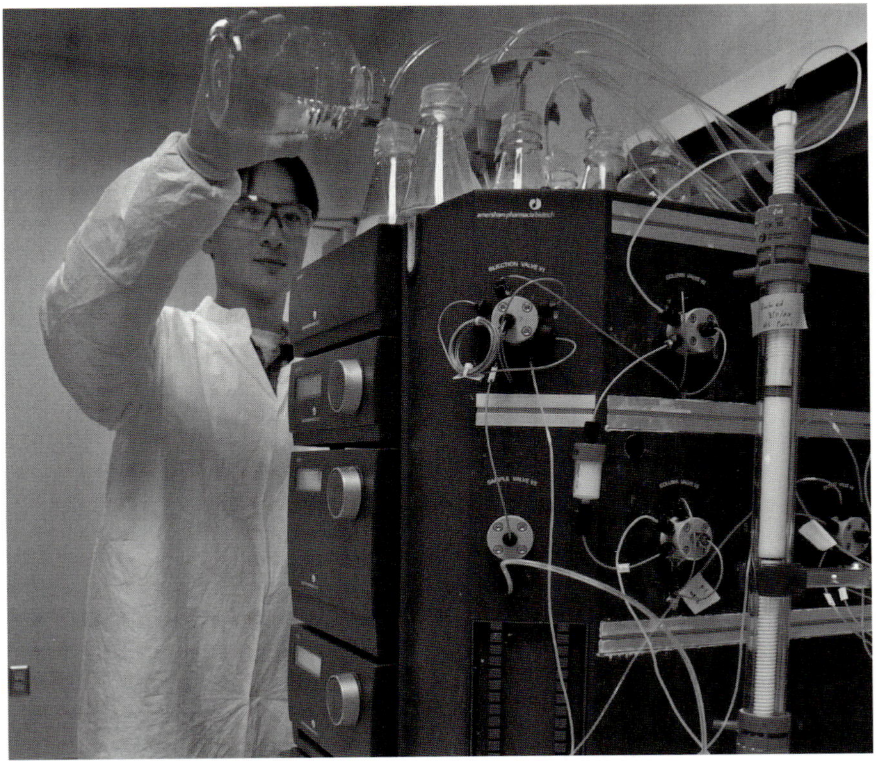

Scientists are pursuing cures for diseases based on gene therapy including PID. Recent experiments helped several children with serious immunodeficiencies who were confined to isolation chambers live more normal lives.

There are more than seventy different types of PIDs. All PIDs result from gene abnormalities. Each type has different symptoms, depending upon which genes are defective or missing. Some of the conditions are mild and cause few problems. Others may be serious and life threatening.

While many PIDs are first detected in infancy, others don't show up until adolescence or adulthood. Symptoms of PIDs include an increased number of infections and health problems that cannot be treated with antibiotics and other conventional cures. Such serious conditions include blood poisoning, *abscesses* (collections of pus in the tissue), and a failure to grow normally. The infections may result in hospitalization.

In the past, babies born with PID were in serious trouble. With some types of PID, the infants could not survive for more than a few months. Today, doctors use antibiotics, gene therapy, and hormone replacement therapy to successfully treat people with many PIDs.

SEVERE COMBINED IMMUNODEFICIENCY DISORDER

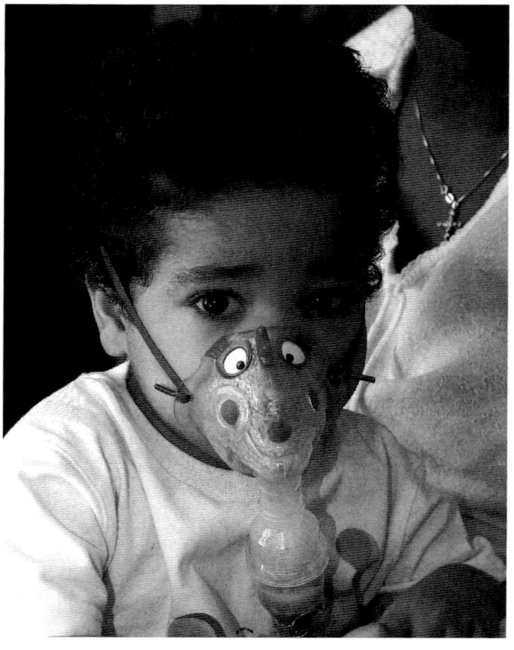

This boy was diagnosed with SCID when he was only four days old. He breathes through a mask to minimize the chances of becoming ill because of airborne bacteria and viruses.

Severe combined immunodeficiency (im-myoon-oh-dih-FISH-en-see) disorder (SCID) is a very rare disease that occurs when a child is born without lymphocytes to fight infection. Without these cells, the entire immune system functions improperly or not at all. Infants and children with this condition suffer frequent, serious, life-threatening infections. Even the most common health problems can be potentially serious for those with SCID.

Symptoms of SCID include failure to thrive, frequent and recurrent ear and respiratory infections that don't respond to treatment, thrush, and diarrhea. Infections may cause organ damage. Most babies are diagnosed with the disease around the age of six months.

In the past, infants with SCID usually died shortly after birth. In the 1960s, doctors began putting children into special isolation chambers, or bubbles, to protect them from disease and infection. In the 1980s, doctors began using stem cell transplants, replacing a patient's bone marrow with a donor's.

More recently, doctors began experimenting with gene therapy to treat and cure the condition. Doctors inject patients with enzymes that are missing in their bodies. This treatment must be continued throughout a patient's life.

A breakthrough came in 2002, when two separate teams of doctors, one in Italy and one in Israel, cured two children born with SCID. Doctors first removed bone marrow from each of the patients. Then they inserted the enzyme-making gene into the marrow stem cells. Finally, doctors transplanted the altered stem cells into the children's bone marrow. Within months, the children's immune systems had become healthy.

The Boy in the Bubble

SCID is sometimes called "the bubble boy disease." It got this nickname from a Houston, Texas, boy named David Vetter. David was born in 1971 with SCID. In order to save his life, his parents had a special plastic isolation chamber, or bubble, built for him. The bubble protected David from the many germs, bacteria, and viruses in the world outside.

David's safety came with a high price. He could never leave his protective bubble, and no one could ever enter. That meant that David could never hug his parents, play with a pet, or run in the park with a friend. As David got older, he began to want a life outside the bubble. So in 1984, the twelve-year-old asked doctors to attempt a then-experimental procedure: a bone-marrow transplant. Unfortunately, complications arose from the procedure, and David did not survive for long afterward. During his last few weeks of life, however, he was able to come out of his bubble and be surrounded by the people who loved him. Thanks to David, public awareness about this serious disease was raised.

David Vetter was born with an inherited disorder that left him no natural immunities against disease. He lived much of his life in a plastic enclosure to protect him from germs, viruses, and bacteria.

Geneticist Yury Verlinsky is a pioneer in the area of prenatal testing for genetic diseases.

DIAGNOSING AND TREATING IMMUNE SYSTEM DISORDERS

A doctor who suspects that a patient is suffering from a PID will first take blood samples. The samples are checked in a laboratory for missing or abnormal genes that might be causing the problems. The doctor may also perform skin tests to see how well the T cells in the body are working. Reaction-provoking substances are injected into the skin. If a person has a healthy immune system, the skin should react within twenty-four to forty-eight hours.

To determine exactly which PID a patient is suffering from, the doctor may next perform sputum tests on the person's mucus. In some cases, doctors may surgically remove a piece of infected tissue for study.

Prenatal testing can detect some PIDs before a child is even born. With this early warning, parents can prepare to care for a baby with special needs. They may also arrange to have a bone marrow transplant performed once the baby is born.

When a patient is suffering from a PID-related infection, the doctor's first concern is to treat the infection. Once that is done, the main goal is to permanently correct the problem with the immune system, if possible. To do this, patients may receive a bone marrow transplant with healthy new cells. These healthy cells will begin to make healthy blood cells and correct problems with the immune system.

Other options for some PIDs are *enzyme replacement therapy, antibody replacement therapy,* and *immunoglobin* (im-myoon-oh-GLAHB-yoo-lin) *therapy.* With these types of treatments, missing enzymes, antibodies, or *globulins* (proteins) are injected into the body. However, patients must receive regular injections to keep conditions under control.

Gene therapy has been tried since 1990. This procedure involves removing T cells from the body, inserting healthy ones into the T cells to replace missing or abnormal genes, and then inserting the cells back into the body. The hope is that the body will begin to correct the genetic defect. More recently, stem cells have been used in gene therapy. This therapy is still considered experimental.

Bone Marrow Transplants

Receiving new bone marrow is similar to receiving blood. Bone marrow from a healthy donor trickles into a vein through an intravenous needle. It circulates through the body and settles in the bone marrow cavities of the body's long bones. Here, it begins producing new cells.

For bone marrow donors, the removal procedure is slightly more complicated. A patient is usually given medicine that makes him or her unconscious. Then doctors insert a long needle into the pelvic bone in the lower back and extract bone marrow.

In order for the bone marrow to work properly, the person donating and the person receiving bone marrow must be a biological match. This means that their bodies' tissues must have nearly identical molecules. If they are not a match, the immune system of the person receiving the bone marrow may attack the new cells. Many bone marrow donors are therefore close relatives of people receiving donations.

Transplant Rejection

Each year, the lives of thousands of Americans are saved by organ transplants. A transplant is a surgical procedure to remove an unhealthy organ or tissue and replace it with a healthy organ or tissue from a donor. In some cases, a living donor chooses to donate an organ that he or she does not need. For example, someone may donate one kidney to help another person. The donor remains healthy, because a person can normally function perfectly well with only one kidney. In other cases, organs for transplanting come from people who are brain-dead—for example, when the body of someone killed in a car accident is artificially kept alive so that doctors can use the organs to save others. This is the case when transplanting vital organs, such as the heart, without which people cannot live.

In some cases, the immune system of the person receiving the new organ goes on the attack. It views the new organ as a foreign invader and sets out to destroy it. This is called *transplant rejection, graft rejection,* or *tissue rejection.* If this condition is not treated immediately and effectively, it can be life threatening.

To cut down on the chances of transplant rejection, tissues from both the patient and the donor are typed. Both sets of tissue are checked for the antigens they contain. No two people will have tissue samples that match perfectly. However, doctors can get a sense of how serious the reaction of the immune system will probably be. Then patients who are going to receive a new organ or tissue are given medicines that suppress their immune system. They begin taking these medicines before surgery and must continue taking them for the rest of their lives.

Even with immuno-suppressant drugs, a person's body may still reject its new organ or tissue. Symptoms of transplant rejection vary, depending upon the organ. They include discomfort and a loss of organ function. Some people may also experience pain in the transplanted organ and fever. If the rejection cannot be controlled, a person may lose all function in the organ, which may be fatal, if the organ is a vital one, like the heart, liver, or kidneys.

9

HOW THE ENVIRONMENT AFFECTS THE IMMUNE SYSTEM

Bacteria, viruses, germs, allergens, air pollutants—these substances are all around us. Humans can't help inhaling, ingesting, and absorbing substances that can cause problems for the immune system. However, other environmental diseases can be avoided.

ALLERGIES

Allergies are an abnormality of the immune system. When a person has allergies, the immune system responds incorrectly to a normally harmless substance, such as dust, pollen, pet dander, or mold. The substance might also be a certain food, chemical, or medicine. Even a bee bite or other insect venom may cause an allergic reaction in some people. Tobacco smoke is a common allergen for many people.

At the start of an allergic reaction, the body's immune system identifies the common substance, called an *allergen,* as a foreign substance that must be destroyed. It begins to manufacture special antibodies to fight the substance. For each different allergen to which a person reacts, the body creates a different, unique antibody.

> **Fast Fact**
>
> One ragweed plant can make a million grains of pollen per day.

When the antibody and the allergen come in contact, the body begins to release substances that cause allergy symptoms. These symptoms include itchy eyes, nose, and throat; coughing; sneezing; dark circles under the eyes (called *allergic shiners*); increased phlegm production; and conjunctivitis. *Conjunctivitis* (kun-junk-tih-VYE-tiss) is an inflammation of the eyelids. It causes the eyes to become red, encrusted, and swollen. Some people with allergies may also develop asthma.

To many people, allergies are uncomfortable and annoying, but for some people, allergies can be life threatening. A severe allergic reaction causes a condition called *anaphylaxis* (an-uh-fih-LAK-suss). Symptoms of this condition include difficulty breathing, swollen tongue, dizziness, and fainting. If not treated, death can result. Substances known to have caused anaphylaxis include foods, latex, insect bites, and medicines.

As many as 50 million Americans suffer from some sort of allergy. Two million of those sufferers are children. Allergies cause a loss of school days, workdays, and billions of dollars in healthcare costs each year.

Some people have *seasonal allergies,* which occur only during certain seasons. Many people who have seasonal allergies suffer during the spring, when trees and flowers pollinate. Others have *perennial allergies.* This means that they are allergic to something year-round. Dust, for example, is present throughout the year.

Many allergies are caused by airborne substances, including pollen, animal dander, mold, and chemicals. Of all airborne allergens, dust mites may be the most common cause of continual allergies.

People can also be allergic to certain foods. Common food allergens include peanuts and other nuts, milk, soy, eggs, and seafood. The first symptoms of a reaction to a particular food might be an itchy mouth and throat. Other symptoms may include cramps, vomiting, diarrhea, rash, and shortness of breath.

Why do some people have allergies? Health officials believe that people may inherit the tendency to be allergic. Allergies often seem to run in families. However, people probably don't inherit allergies to any specific substances.

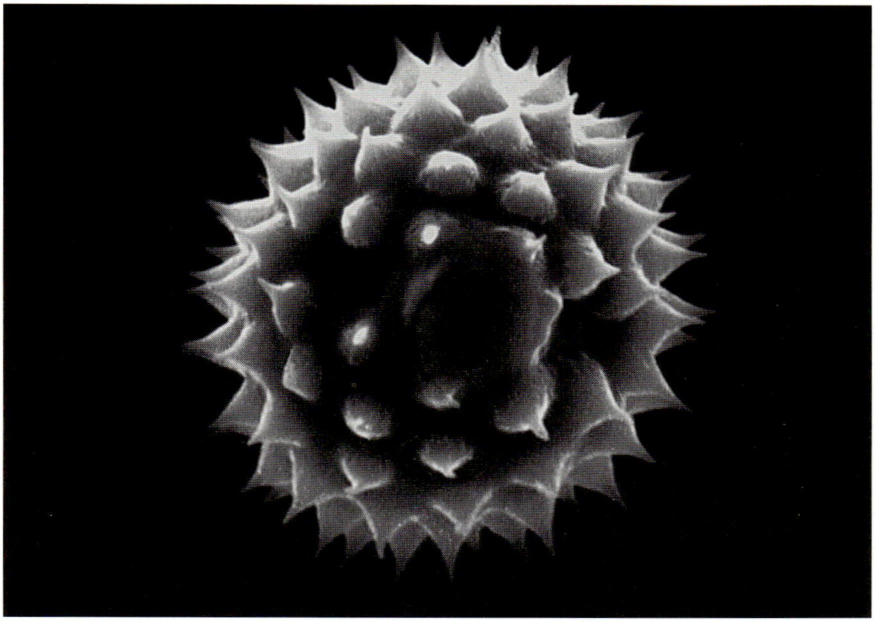

Springtime pollen like the sample shown here causes allergic reactions in millions of people. The image of this pollen sample has been enlarged many times by an electron microscope.

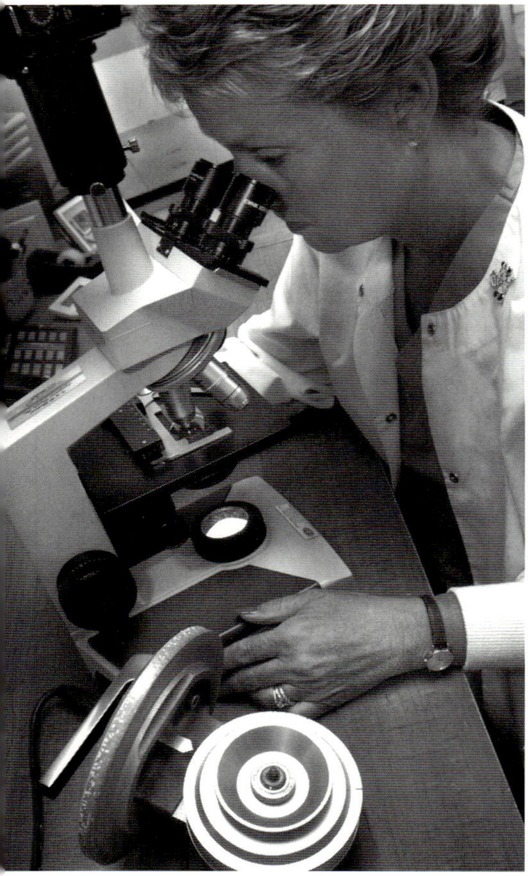

An allergy specialist analyzes pollen and mold under a strong microscope.

DIAGNOSING AND TREATING ALLERGIES

Doctors test for allergies by checking to see which substances a person reacts to. They perform skin tests, pricking the patient on the arm or the back with needles that contain different allergens. If a patient is allergic to one or more of the allergens, a red, itchy bump appears on the skin where the allergen was placed. The larger the bump, the more severe the skin reaction. Sometimes, blood tests are used to determine which allergens a person reacts to.

Once doctors have determined what a person is allergic to, they can use three different ways to treat allergies: avoidance, medication, and immunotherapy. *Avoidance* is the first course of treatment. This is the attempt to eliminate a person's exposure to the allergen that causes problems. In some cases, this is simple. If a person has an allergy to seafood, for example, he or she can simply stop eating it. In many cases, however, avoidance is not possible. Dust, for example, is hard to escape. It is found everywhere. Although there are ways to reduce exposure to dust in the home, it is unlikely that a person with dust allergy will be able to avoid it completely.

When avoidance is not possible, doctors may prescribe *medication* to relieve allergy symptoms. Such medicines may include antihistamines, decongestants, and nasal sprays that contain steroids.

In some cases, doctors may recommend that a patient receive allergy shots, or *immunotherapy.* These shots desensitize a person to the allergen. They contain a purified form of the allergen to which the patient reacts. The allergen will gradually build up in the body, allowing the immune system to become used to the substance. Immunotherapy provides long-lasting, if not permanent, relief to allergy sufferers. People usually receive immunotherapy shots for about five years.

MULTIPLE CHEMICAL SENSITIVITY

Multiple chemical sensitivity (MCS) is a type of allergy in which people react to various chemicals that they come in contact with, such as tobacco smoke, pesticides, perfumes, cleaners, and soaps. MCS may be triggered by one severe exposure to a chemical or other pollutant. It may also be set off by long-term, continuous exposure.

Symptoms of MCS include fatigue, memory loss, headaches, rashes, asthma, and aching muscles and joints. Although some doctors are certain that MCS is a valid condition, others question whether or not it truly exists. They point out that there is no scientific evidence to prove that such a condition as MCS truly affects people.

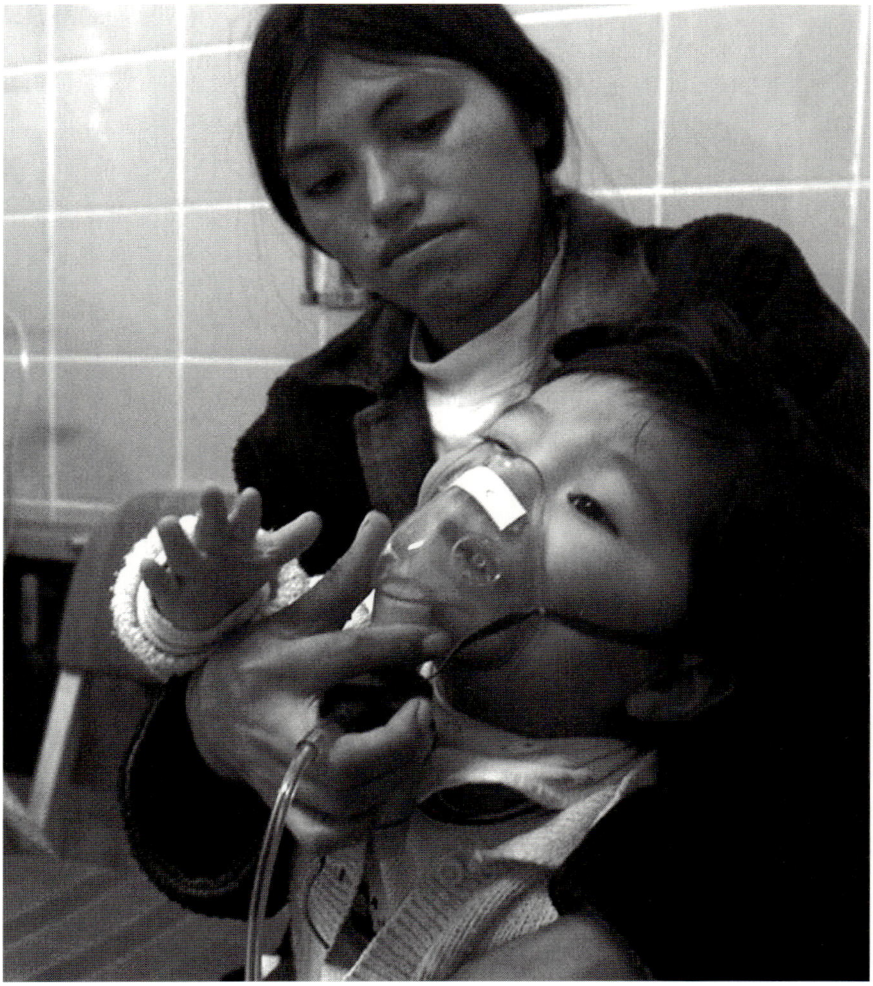

A mother holds a mask over the face of her one-year-old child to clear his lungs from the effects of pollution and high humidity.

SICK SCHOOL SYNDROME

A 1995 study found that half of all school buildings have problems with indoor air pollution. These "sick buildings" can pose a serious health problem for kids with asthma, allergies, and other respiratory problems. Symptoms of this *sick school syndrome* include headache, fatigue, dizziness, nausea, and difficulty breathing.

Many factors can contribute to sick school syndrome. One major factor is poor ventilation. Poorly ventilated areas are perfect for the rapid growth of molds and bacteria. In newer schools that have been built to conserve energy costs, ventilation is especially important. That's because chemicals, cleaning agents, and pollution from synthetic building materials are more likely to remain inside these well-sealed buildings for long periods of time.

Sick school syndrome is just a small part of a larger problem known as *sick building syndrome* (SBS). People affected by SBS suffer from health problems that have no other known cause. In addition, their symptoms disappear when they are out of the "sick building" and reappear when they reenter the building. These health problems affect more than one or two people who work or live in the building.

To keep children safe at school, the Environmental Protection Agency (EPA), a government agency formed to protect and preserve the environment, has developed an Indoor Air Quality Tool for Schools kit. The kit contains information that helps school administrators assess whether their buildings are at risk and take action to make life healthier for students. It includes checklists for school employees, a fact sheet about indoor air pollution, and sample policies and memos that administrators can use.

A student boards a school bus at North Country Union High School in Vermont. In order to reduce air pollution, the school district has strict rules against buses idling too long near school buildings.

AIDS

The *human immunodeficiency virus* (HIV) is a contagious virus that attacks the body's immune system. One of the chief ways that HIV does this is by destroying T cells. HIV can lead to *acquired immune deficiency syndrome* (AIDS), a serious, life-threatening condition.

People with AIDS suffer from a weakened and damaged immune system. Their immune system can no longer fight off other infections and diseases. As a result, they become ill more often than other people and have a more difficult time recovering from each illness. AIDS can lead to cancers and infections of the lungs, skin, brain, and digestive system.

A CONTAGIOUS DISEASE

HIV is spread through contact with infected blood and other body fluids. Contaminated blood and fluids can be transmitted from one person to another in several ways, including:

- unprotected sexual intercourse
- sharing of dirty needles or syringes when using illegal drugs
- blood transfusions
- pregnancy, birth, or breastfeeding (from infected mother to an infant)

It is less likely that people will be infected with HIV by coming in contact with blood from an open wound, using razors or other personal hygiene items, or being bitten by someone who has AIDS.

HIV *cannot* be transmitted from one person to another by:

- using the same toilet, towel, utensils, clothes, or swimming pool
- hugging, holding hands, or casual kissing on the cheek or lips
- sneezing or coughing
- participating in sports or games in which there is no risk of bleeding
- being bitten by a mosquito or other insect that has already bitten an infected person

How Safe Is the Blood Supply?

In the late 1970s and early 1980s, when the HIV virus was first recognized, a number of people contracted the disease when they received contaminated blood during surgical procedures. Today, however, strict rules are in place to protect the blood supply in the United States. Since 1985, all donated blood has been carefully tested for signs of HIV and other diseases. Today, health experts believe that the chance of contracting HIV through a blood transfusion is 1 in 450,000 to 1 in 600,000.

A lab technician fills a tray with blood samples that will be analyzed for various diseases.

SYMPTOMS

Some of the first symptoms of HIV infection are similar to the symptoms of the flu. These include cough, fatigue, fever, headache, and swollen lymph nodes. The symptoms may appear a month or two after a person has been exposed to HIV.

More serious symptoms may not appear until ten to eleven years after initial exposure to the virus. These symptoms include swollen lymph nodes under the arms, severe yeast infections, warts, night sweats, weight loss, and diarrhea.

The speed with which HIV lays waste to the immune system varies from person to person. With some adults, for example, serious symptoms may not appear until many years after the onset of infection. Some people have lived with HIV for more than a decade before experiencing such symptoms. Others begin exhibiting symptoms a few months after infection.

HIV is considered to have progressed to AIDS when the number of certain infection-fighting cells in the blood falls below a specific level. As a result, the patient begins to develop serious opportunistic infections or cancers. *Opportunistic infections* are those that would normally not cause much harm to the human body but which can be deadly to people with weakened immune systems. Opportunistic infections that commonly affect those suffering from AIDS include thrush, pneumonia, fungal infections, viral brain infections, some cancers, eye diseases, skin rashes, and intestinal infections. Common cancers that affect people with AIDS include cervical cancer, Kaposi's sarcoma, and lymphoma.

The most common opportunistic infection that people with AIDS get is *pneumocystis carinii* (noo-moh-SISS-tiss kuh-RYE-nee) *pneumonia* (PCP). In the past, it has been the number one cause of death in people with AIDS. PCP is a respiratory infection caused by a fungus that is present in the human body. As the immune system of a person with AIDS becomes weaker, it can no longer control the fungus. The fungus attacks the lungs, causing pneumonia.

HISTORY OF AN EPIDEMIC

The disease that the world would come to know as AIDS first struck the United States in 1981. At that time, few people knew anything about it. How was it transmitted? Who was at risk for it? Most importantly, how was it to be treated and possibly cured?

> **Fast Fact**
>
> AIDS is the fifth leading cause of death among Americans between the ages of twenty-five and forty-four.

Today, doctors know much more about AIDS and HIV. People know more about behaviors and conditions that might put them at risk. This knowledge and years of research have led to better treatments that have helped people suffering from AIDS live longer, healthier lives.

A grandmother from the African country of Malawi cares for her eleven grandchildren, who were all orphaned by the AIDS epidemic. About 9 percent of Malawi's population of 10 million people are HIV positive.

One question remains unanswered. Right now, there is no cure for HIV or AIDS. Researchers continue to look for the key that will unlock this mystery. They are studying people who have been infected with HIV for more than ten years without ever becoming ill. They are testing new vaccines in the hopes that one will prevent HIV infection. They are looking for new and more effective treatments for the opportunistic infections and other conditions that AIDS patients suffer from, as well.

Since 1981, more than 700,000 cases of AIDS have been reported in the United States. According to the CDC, 40,000 new cases of HIV infection are reported each year. Men account for seven out of ten of all new HIV cases.

The disease has also spread around the globe. As of June 2000, nearly 19 million people had died as a result of AIDS and more than 34 million people were currently living with HIV or AIDS.

That number grows every day. In fact, experts say that AIDS has reached epidemic proportions in some nations.

Africa is one continent that has been devastated by the disease. At the Fourteenth International AIDS Conference, held in July 2002, experts said that AIDS cases in Africa continue to grow, despite earlier predictions that the worst was over. In Botswana, for example, 39 percent of all adults are believed to be HIV positive. Experts also warn that the epidemic could soon spread to Eastern Europe, the Caribbean, India, and China.

AIDS Statistics

In 2002, the WHO announced that the spread of AIDS around the globe had reached a crisis stage. Here's a look at the incidence of AIDS in various parts of the world.

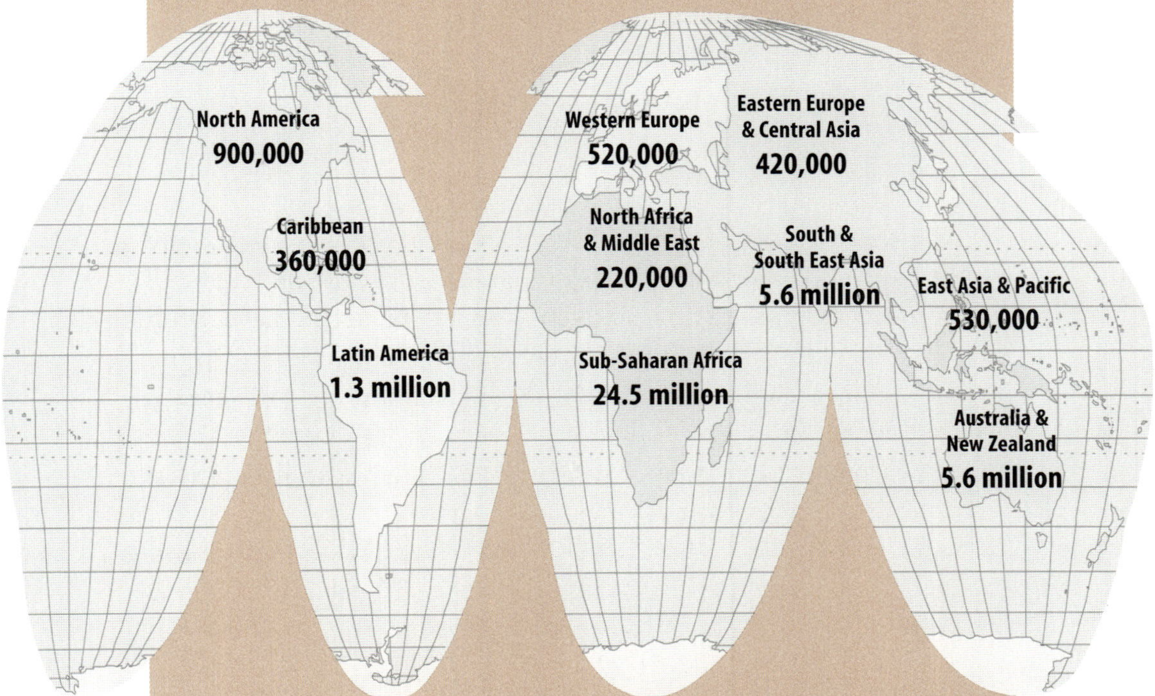

TESTING AND TREATMENT FOR HIV AND AIDS

People who think that they may have contracted HIV can be tested in a number of ways. Blood, mouth cells, and urine can all be checked for the presence of the virus. One common test is the ELISA (enzyme-linked immunoassay) test. If this test is positive, a different test must then be performed. That's because some other medical conditions, including Lyme disease and lupus, may also cause a positive ELISA result. The second test, the Western Blot test, is used to confirm a diagnosis of HIV infection.

Even if people think that they may have been exposed to HIV, the CDC recommends that they wait for three months before being tested. HIV will not show up in blood tests until then. Another test, to confirm results, should be taken six months after the first test.

In the 1980s, a person diagnosed with AIDS had a slim chance of living for more than a few years. Today, thanks to new drugs and a better understanding of the condition, people with AIDS have longer, healthier lives. They can take antiviral medications to prevent such conditions as PCP.

A young woman reacts during an HIV/AIDS test. Early detection of the disease can dramatically improve the chances of survival and lower the risk of spreading the disease to others.

10

KEEPING THE CIRCULATORY, RESPIRATORY, AND IMMUNE SYSTEMS HEALTHY

It's easy to become worried when thinking about all the things that can go wrong with the body. Some health problems, such as congenital conditions, are not in your control. However, the chances of getting many disorders and diseases can be reduced or avoided by practicing good health habits. Regular exercise, a nutritious diet, and other healthy habits will keep your body in top shape. The earlier you start, the better your habits, and the fitter your body will remain.

A six-year-old participates in Project Fit America, which is aimed at introducing children to the benefits of healthy diet and activity.

DIET

Eating healthy foods is one of the best and easiest ways to keep the entire body healthy. A good diet is especially important for a healthy circulatory system. Low-fat diets that are also low in cholesterol keep the heart and blood vessels working properly. A low-fat, low-cholesterol diet also reduces the risk of blood clots and other problems.

People who are concerned about reducing fat in their diets should choose poultry and fish over red meat for many main meals. This is because red meats contain high levels of saturated fats. Saturated fats are considered "bad fats" because they increase the amount of cholesterol in the blood. Diets that are high in sodium can be harmful, too. Although sodium plays an important role in regulating normal heart rhythm, too much sodium causes the body to retain fluids. This puts extra stress and strain on the heart and other organs.

Fast Fact

The iron found in red meats, poultry, and fish is more easily absorbed by the body than the iron found in leafy green vegetables and iron-fortified breads and cereals.

Doctors recommend that people should eat plenty of fresh fruits and vegetables. People who eat at least three servings of fruits and vegetables each day have a much lower risk of dying from stroke, heart diseases, and other causes than people who don't eat these foods. A recent study found that for the fruit- and vegetable-eating group, chances of dying from a stroke were reduced by 42 percent.

A balanced diet helps people control their weight. *Obesity,* the condition of having an excess amount of body fat, is becoming a problem in the United States. Studies report that more than 108 million Americans over the age of twenty are overweight; 44 million of these people are considered obese. Obesity damages the body. It puts extra strain and pressure on the heart and the lungs. It can lead to such health conditions as CAD, stroke, high blood pressure, and high cholesterol levels.

Here's how vitamins affect the circulatory, respiratory, and immune systems.

- *Vitamin K* is important to blood clotting. It can be found in green leafy vegetables, egg yolks, liver, and fish oils. Vitamin K is also made by some types of intestinal bacteria.
- *Vitamin B[1],* also called thiamine, helps the heart function properly. It can be found in dairy products, whole grains, eggs, and beans.
- *Vitamin B[12]* is essential for the production of red blood cells. It can be found in eggs, dairy products, fish, poultry, and meat.
- *Vitamin C* may help the body resist infection and fight wounds better. It is found in citrus fruits (oranges, lemons, limes), tomatoes, potatoes, and strawberries.
- *Vitamin B_9,* also call folic acid, is another nutrient that promotes the production of healthy red blood cells. For this reason, people with sickle-cell anemia often take folic acid supplements. This type of B vitamin can be found in fruits, beans, and vegetables.
- *Potassium* helps regulate the heart's rhythm and may reduce high blood pressure. This element is found in bananas, cantaloupe, grapefruit, tomato juice, potatoes, and prunes.

Dr. Edward A. Doisy won the 1943 Nobel Prize in Medicine for his discovery of the chemical properties of vitamin K. Vitamin K is vital for blood coagulation.

- *Iron* is an essential nutrient that every human body needs. Without it, the blood cannot produce enough hemoglobin in red blood cells. This important mineral can be easily obtained by eating a well-balanced diet that includes iron-fortified breads and cereals, red meats, leafy green vegetables, tomatoes, raisins, and egg yolks.

Good Cholesterol, Bad Cholesterol

Nutritionists sometimes talk about "good cholesterol" and "bad cholesterol." What they're really talking about are *lipoproteins* (LYE-poh-proh-teenz), proteins that carry cholesterol throughout the body.

High-density lipoprotein (HDL) is what is known as good cholesterol. Researchers think that HDL carries cholesterol away from arteries and into the liver, where it is expelled from the body as waste. Only about one-third or one-fourth of all cholesterol is HDL cholesterol.

Low-density lipoprotein (LDL) is what is known as bad cholesterol. LDL carries most cholesterol through the arteries. An excess of this cholesterol can cause plaque buildup in the blood vessels, clogging arteries and leading to heart attacks and other circulatory problems.

UNDERSTANDING CHOLESTEROL

Cholesterol (kuh-LESS-tur-awl) is important to the body. This soft, waxy substance is found in the blood and cells of the body. It helps the body produce cell walls, hormones, and tissues. It also reacts with sunlight to produce vitamin D. The human body produces all the cholesterol it needs. Every day, the liver makes about 1,000 milligrams (0.035 ounces) of cholesterol.

Cholesterol is also found in foods from animals, such as meat, milk, cheese, and eggs. Cholesterol from dietary sources can have a negative impact on health by raising the overall amount of cholesterol in the body. A high level of cholesterol in the blood is called *hypercholesterolemia* (hye-per-kuh-less-tur-awl-EE-mee-uh). It can cause some serious health conditions, such as CAD and heart attacks. It is a major health problem in the United States. More than 102 million Americans have high levels of cholesterol in their blood.

Health officials recommend that people begin limiting their cholesterol levels early in life. Some people inherit high cholesterol levels from their parents. These people especially need to be aware of what they're eating and how cholesterol is impacting their health. Obese people are also likely to have high cholesterol levels.

Some Cholesterol-Cutting Tips

- Know what you're eating! Examine nutritional information labels.
- Limit cholesterol intake to 300 milligrams (0.01 ounces) daily.
- Choose lean meats and low-fat products.
- Remember that fruits, vegetables, and grains have no cholesterol.
- Exercise! It increases the levels of "good" cholesterol in the body.
- Avoid fried foods.

EXERCISE

Exercise is good for you. It keeps the heart healthy, the lungs working well, and the immune system functioning properly. Unfortunately, some Americans are not getting the exercise that they should. In 1998, 29 percent of Americans over the age of eighteen reported no leisure-time physical activity. Inactive people have a 30 to 50 percent higher chance of getting high blood pressure.

Exercise is especially important for a healthy heart. That's because the heart is a muscle. For muscles to remain in tip-top shape, they need to be exercised. Health professionals recommend regular daily exercise, at a level that suits you. Remember to warm up and stretch before you start and don't overdo it.

You should also know that exercise:

- improves the circulation of blood through the body
- lowers blood cholesterol levels
- controls high blood pressure
- loosens mucus in the lungs
- reduces the chances of blood clots and embolisms

People with circulatory and respiratory ailments should always check with their physician before starting any new physical activities.

The Children's Health and Exercise Club is one of the nation's first fitness centers devoted exclusively to children. The exercise equipment has been specially modified to allow even very young children to participate.

CIGARETTES AND OTHER DRUGS

Smoking affects the entire body. It increases the risk of serious health conditions, such as pneumonia, asthma, cancer, and heart disease. If you know people who smoke, encourage them to quit. The chances of getting lung cancer decrease each year that an ex-smoker stays tobacco-free. This is because the abnormal cells in the lung are eventually replaced by new, healthy ones.

Illegal drugs can cause serious damage to the body, too. Cocaine and marijuana, for example, can raise the risk of suffering a heart attack. Injecting illegal drugs into the veins can cause infections. Using a dirty needle can lead to AIDS, tuberculosis, and other diseases.

Cocaine is one drug that has been directly linked to a number of circulatory problems. It can raise blood pressure and cause the walls of the heart to thicken. Cocaine has also been linked to a deadly heart condition called *aortic dissection.* Aortic dissection is a tearing of the wall of the aorta. Other drugs can cause blood vessels in the brain to rupture. These include phencyclidine (fen-SYE-klih-deen), or PCP, and lysergic acid diethylamide (lye-SER-jik ASS-id dye-ETH-ul-uh-myde), or LSD.

Alcohol is a drug, too. Like other drugs, it can harm the body. Alcohol elevates blood pressure, raises the amount of fats in the blood, and increases the risk of heart disease and heart failure. Excessive and prolonged alcohol intake can destroy heart muscle tissue.

People who use drugs are also more likely to engage in other unsafe behaviors. In many cases, the drugs lower inhibitions, making people more likely to do things that they might not normally do. This includes such risky behavior as reckless driving and unprotected sexual intercourse. Unprotected sex can lead to HIV infection, sexually transmitted diseases, and unwanted pregnancies.

STRESS

Research has shown that people who habitually become tense, angry, and stressed-out when under pressure are more likely to have heart attacks than their easygoing friends. Doctors say that stress causes the blood vessels to constrict, affecting the flow of blood. New studies have shown that chronic stress may also affect the functioning of the immune system.

People with heart disease and other health conditions must find ways to reduce stress. It is important for them—and everyone—to learn stress relief techniques so they can calm down when things get hectic.

How Can You Fight Stress?

- Stay positive.
- Exercise regularly.
- Talk to friends, family, and other people whom you trust when things get rough.
- Take a few moments each day to just sit, relax, and clear your mind.
- Try to avoid upsetting situations that are not necessary to deal with. If you must face an upsetting situation, try to think of positive solutions.

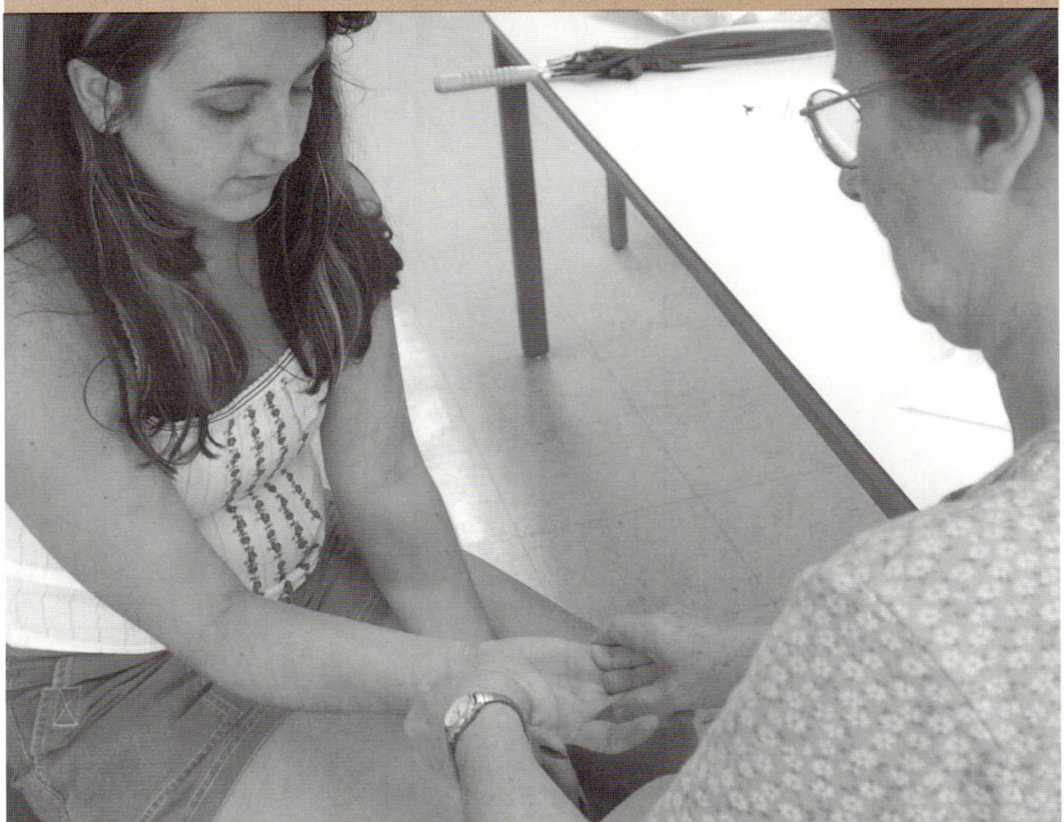

A young flood victim has her hands massaged during the Mcdowell County Healing Arts Day in West Virginia. The event was held to relieve stress for flood victims in 2002.

IMMUNIZATION

Prevention of disease is the best way to control public health. An important part of prevention is immunization. *Immunizations,* also called vaccinations, are shots that contain vaccines—weakened or dead bacteria or viruses that cause various diseases. A vaccine causes the immune system to make antibodies that protect the human body from those diseases.

Before immunizations were available, such diseases as smallpox, polio, and measles were serious, disabling, and often deadly. Many of these diseases struck children. Today, many of these terrible childhood scourges can be avoided with vaccinations.

Children should get certain immunizations at specific ages. Vaccinations begin when a child is about two months old. Most end by the time the child is six years old. Immunizations protect children against polio, rubella, mumps, measles, chicken pox, and many other conditions. These vaccines ensure that children will grow up as healthy and disease-free as possible.

Fast Fact

Did you know that houseplants can make some people ill? That's because damp soil in a plant's pot can become moldy, releasing mold spores into the air. People with allergies should avoid overwatering their plants or try enclosed, terrarium-type plants instead.

Despite the benefits of immunizations, the CDC estimates that as many as 56 percent of all American children are not fully immunized by the age of two. If children do not receive the proper vaccines, they may not be allowed to take part in school, camps, sports, and other activities.

Health officials say that it is important for adults to receive vaccinations, too. Recommended immunizations protect against flu, hepatitis (hep-uh-TYE-tiss) B, pneumococcal infections, tetanus, and diphtheria. It is especially important for elderly people and those suffering from certain health conditions, such as heart conditions and weakened immune systems, to stay up to date on their immunizations. Health experts suggest that senior citizens receive vaccinations against flu, pneumonia, tetanus, diphtheria, and other conditions.

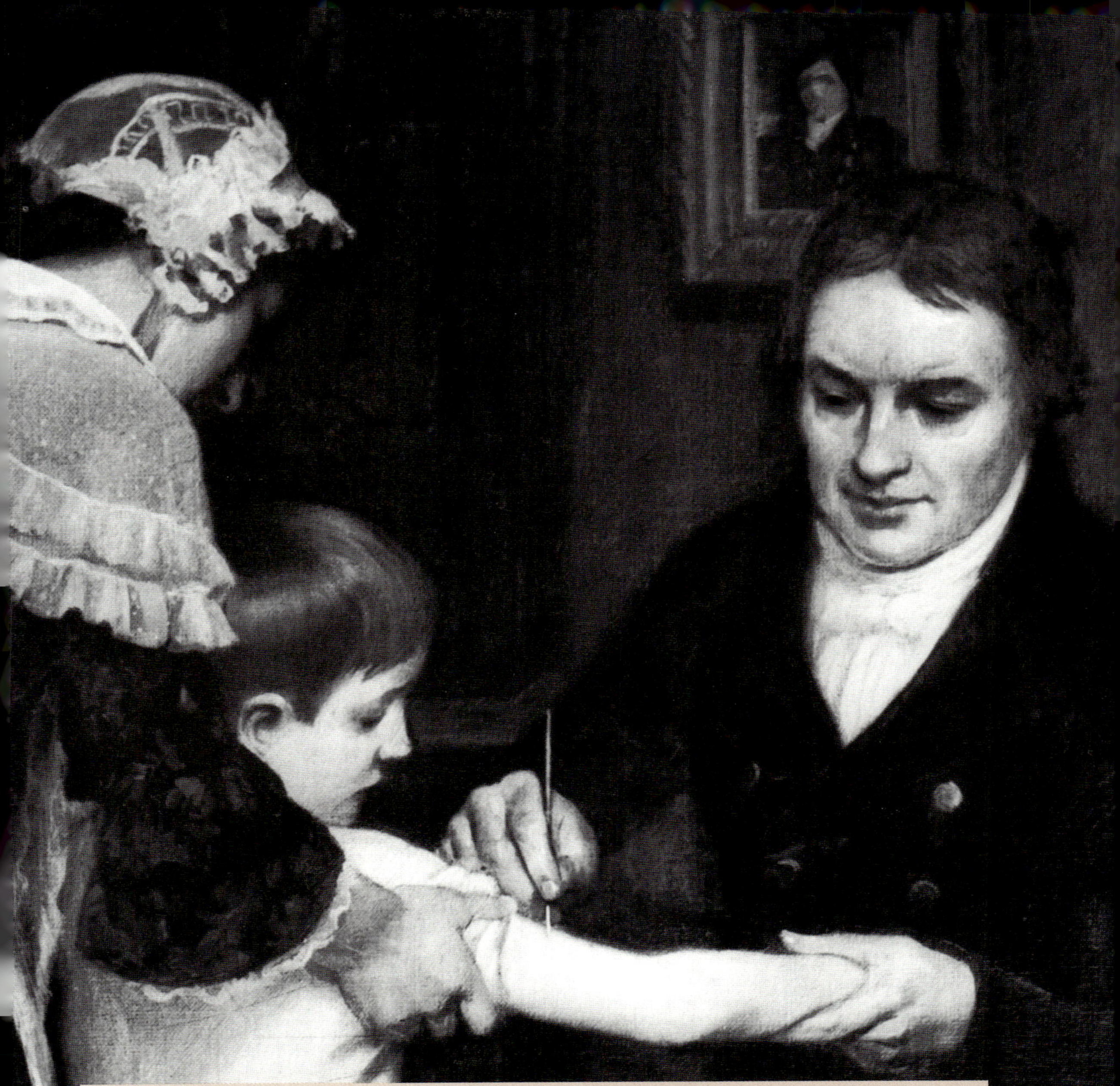

Health and History: The First Vaccine

The first vaccine was developed in 1796 by English physician Edward Jenner (1749–1823). Jenner noticed that milkmaids who suffered from a common disease called *cowpox* did not suffer from the disease *smallpox.* At this time, smallpox was one of the most feared diseases in England. It caused permanent scarring, blindness, infertility, and death.

Jenner drew some pus from the cowpox sore of a milkmaid. He then injected the fluid into the arm of a healthy eight-year-old boy. A few weeks later, he injected the same boy with fluid from a smallpox sore. To many people's surprise, the boy remained healthy. Jenner had performed a medical miracle.

INDOOR AIR POLLUTION

People with respiratory problems should do their best to reduce the amount of pollution that they breathe while indoors. Here are a few tips:

- Encourage family members to make your home a "no smoking zone."
- Make sure that your home, especially the kitchen, is well ventilated.
- Check for plywood, fiberboard, or particleboard furniture that may be coated with formaldehyde.
- Cut down on dust and dirt in the home by frequently changing air filters in air conditioners and vacuums.
- Avoid using aerosol dusting powders and other products.
- Try to use natural, chemical-free cleaners for cleaning jobs around the house. Vinegar and baking soda are two effective cleaning tools.
- Check for asbestos and radon frequently.

Don't Spread Disease

Here are five simple ways to avoid a viral or bacterial infection, according to the Association for Professionals in Infection Control and Epidemiology:

1. Wash your hands frequently. This is especially important before handling food and after using the bathroom.
2. Take all medications exactly as prescribed. Be sure to finish the entire course of medicine.
3. Keep immunizations up to date.
4. If you have symptoms of a flu or other infection, don't go to school. Keep your germs at home.
5. Don't share cups, utensils, toothbrushes, drinking glasses, or any other personal item with others.

GLOSSARY

acquired immune deficiency syndrome (AIDS)—a serious, life-threatening condition; people with AIDS suffer from a weakened and damaged immune system, which makes them susceptible to many other diseases

allergy—an abnormality of the body in which the immune system reacts to a normally harmless substance by producing antibodies to attack the substance

arteries—strong, elastic blood vessels that carry blood away from the heart

asthma—an inflammation and narrowing of the bronchial tubes, or air passages, in the throat

autoimmune disorder—a condition caused when the body's immune system attacks other parts of the body that it mistakenly identifies as foreign or harmful

bacteria—microscopic, single-celled organisms that are found everywhere; many bacteria cause disease, but not all are harmful

biopsy—a surgical procedure to remove tissue, cells, or fluids from the body for examination

birth defect—any physical problem with which an infant is born

blood pressure—a measurement that tells how much force, or pressure, blood is exerting against the walls of blood vessels

bone marrow—a type of tissue found in the center of many long bones where infection-fighting cells are produced.

cancer—a disease in which a group of abnormal cells in the body divide without control

capillaries—tiny blood vessels that link arteries and veins

cardiovascular disease (CVD)—any heart disorder caused by a reduced supply of blood to the heart; also known as *heart disease*; CVDs include coronary artery disease, heart attack, and stroke

chemotherapy—the use of anticancer drugs to cure or stop the spread of cancers

cholesterol—a soft, waxy substance that is found in the blood and cells of the body

circulatory system—the body system responsible for circulating blood, oxygen, and nutrients throughout the body

computerized tomography (CT)—a test that uses a thin beam of electromagnetic radiation to create a three-dimensional picture of what's inside the body

cystic fibrosis (CF)—a congenital disease in which mucous membranes in the respiratory tract produce abnormally thick mucus

environmental health—the body's reaction to conditions that humans have little or no control over, including sunlight, bacteria, chemicals, and pollution; it may also include medical problems that result from personal lifestyle choices, such as smoking

fungi—parasitic organisms, such as mushrooms and molds, that feed on both living and decaying matter

genetic disorder—an abnormality in a person's genes or chromosomes

hormone replacement therapy—a treatment in which patients are given artificial hormones to replace those that they are lacking

human immunodeficiency virus (HIV)—a contagious virus that attack the body's immune system; HIV can lead to AIDS

immune system—the body system made up of organs, cells, and special circulatory vessels that perform the job of fighting foreign invaders in the body

leukemia—a cancer of the blood and blood-forming tissues

lower respiratory tract—the part of the respiratory system that is made up of the larynx, trachea, and lungs

magnetic resonance imaging (MRI)—a test that uses magnets and radio waves to create an image of the inside of the body

malignant—cancerous or harmful

myocardium—the muscle that makes up the walls of the heart's chambers

occupational lung disease—a lung condition caused by long-term exposure to irritating or toxic substances in the workplace

plasma—the clear liquid part of blood that carries blood cells and contains dissolved nutrients, minerals, and proteins

platelets—blood cells that contribute to clotting

pneumonia—a viral infection of the lower respiratory tract

radiation—energy in the form of rays or particles

red blood cells—cells that carry oxygen to and collect carbon dioxide and other gases from the body tissues

respiratory system—the body system that controls the inhalation of oxygen and the exhalation of waste gases

stroke—a condition in which the blood supply to the brain is cut off and brain cells and tissue begin to die

tumor—an abnormal mass of tissue that grows in the body; tumors may be benign or malignant

upper respiratory tract—the part of the respiratory system that is made up of the nose, nasal passages, and upper pharynx

veins—blood vessels that carry blood from the body to the heart

viral upper respiratory infections—illnesses caused by viruses that affect the upper airways; they include colds and the flu

virus—a simple germ organism that is always harmful to humans; viruses are found everywhere

white blood cells—cells that fight infections and germs

X-ray—a test that uses electromagnetic radiation to take a two-dimensional picture of what's inside the body

BIBLIOGRAPHY

BOOKS

Aaseng, Nathan. *Autoimmune Diseases.* New York: Franklin Watts, 1995.

Clayman, Charles, ed. *The American Medical Association Encyclopedia of Medicine.* New York: Random House, 1989.

———. *The Human Body: An Illustrated Guide to Its Structure, Function, and Disorders.* New York: DK Publishing, 1995.

Edelson, Edward. *The Immune System.* Philadelphia: Chelsea House, 1999.

Johansson, Philip. *Heart Disease.* Springfield, NJ: Enslow Publishers, 1998.

Kittredge, Mary. *The Human Body: An Overview.* Philadelphia: Chelsea House, 2001.

———. *The Respiratory System.* Philadelphia: Chelsea House, 1999.

Mehler, Robert E. *How the Circulatory System Works.* Malden, MA: Blackwell Science, 2001.

Walker, Richard, ed. *Encyclopedia of the Human Body.* New York: DK Publishing, 2002.

WEB SITES

American Heart Association (AHA)
www.americanheart.org

American Lung Association (ALA)
www.lungusa.org

CDC National Center for Environmental Health (NCEH)
www.cdc.gov/nceh

CDC National Center for Infectious Diseases (NCID)
www.cdc.gov/ncidod

The Children's Environmental Health Network
www.cehn.org/cehn

KidsHealth
kidshealth.org

National Lung Health Education Program (NLHEP)
www.nlhep.org

NIH National Heart, Lung, and Blood Institute (NHLBI)
www.nhlbi.nih.gov

NIH National Institute of Allergy and Infectious Diseases (NIAID)
www.niaid.nih.gov

NIH National Institute of Neurological Disorders and Stroke (NINDS)
www.ninds.nih.gov

United States National Library of Medicine
medlineplus.nlm.nih.gov

INDEX

Note: Page numbers in *italics* refer to maps, illustrations, or photographs.